SpringerBriefs in Regional Science

Series Editors

Henk Folmer, University of Groningen, Groningen, The Netherlands

Mark Partridge, Ohio State University, Columbus, USA

Daniel P. McMillen, University of Illinois, Urbana, USA

Andrés Rodríguez-Pose, London School of Economics, London, UK

Henry W. C. Yeung, Department of Geography, Chinese University of Hong Kong, Hong Kong, China

SpringerBriefs present concise summaries of cutting-edge research and practical applications across a wide spectrum of fields. Featuring compact, authored volumes of 50 to 125 pages, the series covers a range of content from professional to academic. SpringerBriefs in Regional Science showcase emerging theory, empirical research and practical application, lecture notes and reviews in spatial and regional science from a global author community.

All titles in this book series are peer-reviewed. This series is indexed in SCOPUS.

Pedro Fierro · Patricio Aroca · Patricio Navia

Abandoned Places in the Digital Era

Spatial Roots of Disaffection
and the Internet's Role in Inclusion

Pedro Fierro
Adolfo Ibáñez University
Viña del Mar, Chile

Patricio Aroca
Universidad Andrés Bello
Viña del Mar, Chile

Patricio Navia
New York University
New York, NY, USA

ISSN 2192-0427　　　　　　　ISSN 2192-0435　(electronic)
SpringerBriefs in Regional Science
ISBN 978-3-031-81872-1　　　ISBN 978-3-031-81873-8　(eBook)
https://doi.org/10.1007/978-3-031-81873-8

© The Editor(s) (if applicable) and The Author(s), under exclusive license to Springer Nature Switzerland AG 2025

This work is subject to copyright. All rights are solely and exclusively licensed by the Publisher, whether the whole or part of the material is concerned, specifically the rights of translation, reprinting, reuse of illustrations, recitation, broadcasting, reproduction on microfilms or in any other physical way, and transmission or information storage and retrieval, electronic adaptation, computer software, or by similar or dissimilar methodology now known or hereafter developed.
The use of general descriptive names, registered names, trademarks, service marks, etc. in this publication does not imply, even in the absence of a specific statement, that such names are exempt from the relevant protective laws and regulations and therefore free for general use.
The publisher, the authors and the editors are safe to assume that the advice and information in this book are believed to be true and accurate at the date of publication. Neither the publisher nor the authors or the editors give a warranty, expressed or implied, with respect to the material contained herein or for any errors or omissions that may have been made. The publisher remains neutral with regard to jurisdictional claims in published maps and institutional affiliations.

This Springer imprint is published by the registered company Springer Nature Switzerland AG
The registered company address is: Gewerbestrasse 11, 6330 Cham, Switzerland

If disposing of this product, please recycle the paper.

To: Lourdes, Violeta, and Daniel

Preface

Navigating the intersecting domains of regional science, sociology, political science, and communications, this book embodies a multidisciplinary approach that offers valuable insights for a diverse audience. We anticipate that its contents will illuminate students, both at the undergraduate and postgraduate levels, and researchers working within these fields. However, its relevance extends beyond the ivory towers of academia: anyone seeking to understand the dynamics of discontent, both in online and offline engagement, as well as the attitudinal implications of territorial inequalities, will find our research pertinent.

Professionals and decision-makers operating on both local and global scales stand to gain unique insights from our work. Our analysis is particularly timely for Chile—our case of study—a nation currently at a critical juncture in its political crisis where the blueprint for future growth is under intense debate. The book's focus on the territorial dimension brings a fresh perspective to these discussions.

Globally, our work holds appeal for three main reasons. Firstly, it investigates the "geography of discontent" in a setting often overlooked in current discourse—Latin America. Due to understandable reasons, such as the availability of data and recent political phenomena and electoral outcomes, most existing studies are confined to the context of consolidated democracies, primarily in Europe and the United States. By spotlighting an emerging democracy in Latin America currently undergoing significant political transformation, our work enriches the discourse with a Southern Hemisphere perspective.

Secondly, our analysis focuses primarily on the attitudinal dimension of discontent, which tends to be less studied in the literature on spatial inequalities and their political outcomes. Most studies are dedicated to understanding electoral results, such as the success of anti-elitist or anti-establishment narratives in specific context. Scholars often assume the presence of sentiments such as anger, abandonment, and alienation, but these feelings are rarely explored empirically. Our proposal aims to delve into classic literature on political science to revisit concepts that will help us assess these underlying elements. In this context, our book will identify ways to measure the attitudinal dimensions that appear to be related to significant contributions to the geography of discontent. This includes not only theoretical elements but also methodological decisions, which we believe are an essential part of our contribution.

Thirdly, our research seeks to explore the democratic implications of new digital platforms against the backdrop of political and social centralism. This is an important subject, as the digital divide is typically studied in terms of interpersonal inequalities. While our study acknowledges these significant contributions, it also offers a reframing of the question by examining the Internet and digital platforms as potential tools for those who are spatially segregated and face geographical obstacles—due to centralism or physical distance—to engage in traditional forms of civic participation. We argue that this territorial dimension of digital democracy remains under-explored in the literature on the Internet and civic engagement. This insight is especially pertinent for policymakers tasked with shaping digital strategies in countries worldwide.

In summary, this book does not seek to reinvent the wheel; rather, it aims to integrate various contributions that, at times, appear insufficiently connected. Our study, based on several articles published in different journals over the past years, endeavors to offer a comprehensive integration of these contributions. Essentially, this book provides a multidisciplinary approach to the phenomenon under study, offering fresh perspectives that contribute to our understanding of how territorial factors shape feelings of abandonment and frustration. Whether viewed through the lens of academia or policy-making, our book illuminates the complex interplay of geography, politics, and digital democracy.

This interdisciplinary work stands to benefit students, researchers, and practitioners. It will also resonate with those seeking to comprehend the nuances of discontent and the repercussions of territorial disparities. Additionally, this book holds profound implications for policymakers. In Chile's context of constitutional reforms, our insights offer a territorially focused lens to enrich the dialogue. Globally, our focus on an emerging Latin American democracy and the digital dimension of democracy in territorially centralized settings is both timely and novel. In essence, our endeavor bridges an academic gap, illuminating the profound interplay between territory, abandonment, and political disillusionment.

London, UK	Pedro Fierro
Viña del Mar, Chile	Patricio Aroca
New York, USA	Patricio Navia
June 2024	

Acknowledgements

Pedro Fierro acknowledges Fundación P!ensa, Universidad Adolfo Ibáñez, and the Cañada Blanch Centre at the London School of Economics and Political Science. Patricio Aroca acknowledges Universidad Andrés Bello, especially CIUDHAD and LETS, at the Faculty of Economics and Business. Patricio Navia acknowledges Universidad Diego Portales and the New York University. All these organizations provided a vibrant community that helped us refine our ideas and develop this work.

We would especially like to acknowledge the donors and the executive committee of the P!ensa Foundation. Their dedication to developing their public opinion project made this book possible, and we are confident that their work will continue to generate important contributions in the future. We also extend our gratitude to various scholars who participated in related projects with us, including Felipe Sánchez-Barría (Pontificia Universidad Católica de Chile), Ignacio Aravena-González, Daniel Brieba, Andrés Rodríguez-Pose, and Ellen Helsper (all of them from The London School of Economics and Political Science), Sebastián Rivera (Universidad Mayor), Francisco Rowe (University of Liverpool), and Camila Quezada and Eduardo Mundt (Fundación P!ensa). Their invaluable contributions helped us rethink our proposal.

Additionally, we extend our gratitude to the scholars who provided invaluable insights during the review processes in various journals, and to our colleagues who shared important comments and suggestions at different conferences where we presented these ideas, such as the World Association of Public Opinion Research (WAPOR), the European Regional Science Association (ERSA), the British and Irish Section of ERSA, Regional Studies Association (RSA), and others.

This work was supported by the Chilean FONDECYT project # 1231927, FONDAP # 15130009, and the ANID-Millennium Science Initiative Program # NCS 2021_063.

Contents

1 Introduction 1
  1.1 Purpose of the Book 1
  1.2 The Challenges 2
  1.3 Case Study: Chile (2017–2023) 3
      1.3.1 Social Outburst and Constituent Process 3
      1.3.2 Centralism, Marginalization, and Territorial Concentration 4
      1.3.3 The Chilean Region of Valparaíso and its Territorial Challenges 6
  1.4 Structure of the Book 7
  References 7

2 Theoretical Framework 9
  2.1 Political Discontent: Specificity, Trajectory, and Discontent 9
      2.1.1 Political Discontent as a Multidisciplinary Phenomenon 9
      2.1.2 Political Discontent in Latin America 12
  2.2 The Unique Problem of Political Disaffection 14
      2.2.1 How to Frame Political Disaffection 14
      2.2.2 The Personal Belief of Self-competence 16
      2.2.3 The Personal Belief of System Responsiveness 19
      2.2.4 Back to the Classics: The Gamson's Hypothesis 21
  2.3 Digital Platforms and Their Democratic Role 22
      2.3.1 Internet's Emergence: First Glances at Its Democratic Impact 22
      2.3.2 Political Efficacy and Digital Platforms 25
      2.3.3 The Concept of Online Political Efficacy 26
  2.4 Territory and Political Engagement 27
      2.4.1 Political Engagement and Space: Some Initial Considerations 27
      2.4.2 The Role of Territory on Civic Participation Patterns 29

		2.4.3	Back to the Classics: The Politics of Resentment and Its Territorial Roots	29
		2.4.4	Geography of Discontent: Political, Historical, Geographical, and Economic Contexts	31
		2.4.5	Concluding Observations	33
	References			33
3	**Methodological Approach**			41
	3.1	Questionnaire Design		41
		3.1.1	Assessing Political Attitudes Usually Associated with Political Disaffection	41
		3.1.2	Internet and Digital Platforms in Integrating Marginalized Areas	44
		3.1.3	Additional Remarks	45
	3.2	The Analysis		46
		3.2.1	Data Sources and Collection	46
		3.2.2	Consistency Analysis	47
		3.2.3	Structural Equation Models (SEM)	49
		3.2.4	Considering the Territorial Context	52
	References			59
4	**Results and Discussion**			61
	4.1	Political Attitudes and Digital Platforms		61
	4.2	Territorial Perspective on the Digital Divide		64
	4.3	Places That Don't Matter and the Feeling of Abandonment		66
	4.4	Preliminary Results About Resentment and Frustration		67
	References			70
5	**Conclusion**			71
	5.1	Final Remarks		71
		5.1.1	Conclusion	71
		5.1.2	Discussion	73
		5.1.3	Limitations and Further Research	74
	References			76
Appendix				77

About the Authors

Pedro Fierro is an Assistant Professor at the Business School at Adolfo Ibanez University (Chile) and a Visiting Fellow at the London School of Economics and Political Science (UK). He is also a researcher at P!ensa Foundation and an adjunct researcher at the Millennium Nucleus for the Study of Politics, Public Opinion and Media in Chile. Recently, he conducted research as a Miguel-Dols Fellow at the London School of Economics and Political Science. Before he completed his Ph.D. in Communication from the University of Navarra, he was a visiting student at the Department of Public Communication at Pompeu Fabra University. His research is focused on Political Communication and Political Geography. He has published articles in different journals, such as Technology in Society, New Media and Society, Social Science Computer Review, and the Cambridge Journal of Regions, Economy and Society.

Patricio Aroca is a Professor at the Facultad de Economía y Negocios at Universidad Andrés Bello, in Chile. He is also a Visiting Associate Professor at the University of Illinois Urban-Champaign. Patricio earned a B.S. in Business (1983) from the Universidad Austral (Chile), an M.A. in Economics (1987) from the Universidad de Chile, an M.Sc. in Policy Economics (1994), and a Ph.D. in Economics (1995) from the University of Illinois Urbana-Champaign. He specializes in regional economics, econometrics, and natural resource economics and, more specifically, works on regional growth and labor interregional migration and utilizes spatial econometrics and input-output analysis. Patricio is the principal investigator of the research nucleus "Regional Science & Public Policy" of Chile's Millennium Science Initiative, and of the research project "Measuring the Impact and Spillover of Chilean Regional Investment" funded by Chile's FONDEF-CONICYT. Patricio has published in numerous international journals and books and has consulted for the World Bank, UNCTAD, IADB and CELADE-ECLAC. Finally, he is currently the elected President of PRSCO and a Board Member of RSAI and RSAmericas.

Patricio Navia is a (Full) Professor of Political Science at the Universidad Diego Portales in Chile and a (Full) Clinical Professor of Liberal Studies at New York University. He has a Ph.D. in Politics from New York University (2003) and has written extensively on democratization, elections, and public opinion.

Chapter 1
Introduction

Abstract This book contributes conceptually and methodologically to understanding two primary phenomena: the contextual and spatial elements that explain citizens' anger, frustration, and sense of abandonment, and the role of digital platforms in including politically marginalized areas. It aims to transition from the geography of voting to the geography of discontent, reinterpreting digital inequalities by integrating the territorial context. The study faces significant challenges, particularly in addressing "left-behind" places. While trajectories of decline and deindustrialization apply to developed countries, they do not necessarily translate to other contexts. In developing countries, abandonment responds to different political and cultural phenomena. This book addresses these complexities while acknowledging the limitations and possibilities for further research. Focusing on the Valparaíso region in Chile, this book applies its proposed framework and methodology. Chile offers an intriguing context with its strong party system, solid institutions, and high Internet penetration, which has influenced civic engagement. Recently, Chile has faced a significant political crisis marked by widespread dissatisfaction and perceived lack of representativeness. The data used pertains specifically to Valparaíso, which, while representative of national realities, also possesses unique characteristics, such as hosting the National Congress and being a focal point of the 2019 social uprising.

Keywords Geography of discontent · Political attitudes · Chile · Territorial inequalities · Digital inequalities

1.1 Purpose of the Book

This book aims to contribute conceptually and methodologically to understanding two primary phenomena. The first is to identify the contextual and spatial elements that can explain the anger, frustration, and sense of abandonment experienced by citizens. Essentially, our proposal focuses on the study of feelings, attitudes, and emotions, seeking to transition from the geography of voting to the geography of discontent. The second phenomenon explores the role that digital platforms could play in including inhabitants of politically marginalized areas. In other words, it reinterprets classic studies on digital inequalities through a spatial lens.

Our approach is informed by our experiences over the past few years, sharing theoretical, methodological, and practical challenges and insights.

This book delves into the intricate relationship between spatial inequalities and political discontent in the digital era. The recent surges in populist narratives, Euroscepticism, and heightened polarization in established democracies underscore the necessity of examining discontent. While many studies concentrate on individual-level explanations rooted in social, economic, or demographic factors, there is an emerging emphasis on the territorial aspect. Landmark events such as Brexit and the rise of Donald Trump nearly a decade ago, alongside recent expansions of Eurosceptic and nationalist narratives in the European Parliament, suggest a spatial dimension that is often overlooked. Certain areas have emerged as strongholds of anti-establishment sentiments, and they are not necessarily rural, poor, or materially deprived.

In this scenario, several contributions suggest that spatial inequalities, economic decline, and deindustrialization processes are intricately linked with support for disruptive narratives. This book endeavors to offer a comprehensive overview of this burgeoning discourse in the context of the Global South, specifically focusing our analyses on the Chilean Region of Valparaíso, the second most important region of the country.

We analyze how spatial marginalization evokes sentiments of abandonment and disillusionment, even before these feelings manifest in voting behaviors. The book provides a panoramic view of discontent in emerging democracies, explores political disaffection as a subset of discontent, and dissects the role of territory and digital mediums in fostering civic engagement. It also elucidates the methodological tools and techniques employed in our study.

To provide empirical depth, the insights in this book draw extensively from surveys conducted in Chile between 2017 and 2023. Chile, with its inherent political and social centralism and recent political upheavals, renders itself a compelling case study. Given that most discourse on the "geography of discontent" centers on Europe and the U.S., our focus on Chile introduces a fresh perspective from the Southern Hemisphere.

1.2 The Challenges

The study presents significant challenges, which are discussed throughout the book. Perhaps the most decisive challenge relates to how we approach these "left-behind" places. The trajectories of decline and deindustrialization may apply to developed countries, but not necessarily to other contexts. This is evident in the recent literature on regional development traps and how these traps can lead to anti-establishment narratives. In developing countries, there are no traditional development traps, but this does not mean that there are no places left behind. Abandonment in these areas seems to respond to different political and cultural phenomena that deserve study, which this book aims to address while acknowledging the limitations and possibilities for further research.

In this context, our goal is to offer a perspective that incorporates the classic and pivotal contributions of economic geography and political science. Additionally, we aim to address this issue from the standpoint of communications, new digital platforms, and new mechanisms of political participation, particularly through the development of what is commonly understood as the digital and democratic divide.

1.3 Case Study: Chile (2017–2023)

To apply the proposed framework and methodology, this book focuses on the specific case of the Valparaíso region in Chile.

First and foremost, Chile presents an intriguing context for such studies. As a Latin American country, it has been notable over the past few decades for its strong party system, solid institutions, rule of law, sporadic corruption, and economic growth (Mainwaring & Scully, 2008; Roberts, 2015). Additionally, it is a country with high Internet penetration (CEPAL, 2016; Subtel, 2017), which has impacted civic engagement for the last decade (Valenzuela et al., 2012, 2014) and continues to pose various digitalization challenges (OECD, 2020).

However, in recent years, Chile has been undergoing one of the most significant political crises since its return to democracy, marked by widespread dissatisfaction and a perceived lack of representativeness (Castiglioni, 2014; Castiglioni & Kaltwasser, 2016; Gamboa & Segovia, 2016; Luna, 2016; Segovia & Gamboa, 2012).

1.3.1 Social Outburst and Constituent Process

In October 2019, Chile experienced an unprecedented social outburst in its recent history, with over a million people protesting in the capital city, Santiago, and an estimated two million across the rest of the country (Somma et al., 2021). This social mobilization was accompanied by widespread destruction and violence throughout Chile, leading to the paralysis of basic services (such as the public transportation system) and a climate of high political polarization and tension.

For some researchers, Chile's social unrest is emblematic because it occurred in a country previously seen—as stated—as a model of democratic stability, economic growth, and social peace, particularly within the Latin American region. In this context, the authors suggest that the political unrest can be explained by inequality. Although Chile had successfully reduced poverty in recent decades, interpersonal inequalities had not diminished at the same pace, leading to dissatisfaction among a populace that had already been voicing concerns about a political representation crisis for quite some time (Morales Quiroga, 2021).

Other scholars offer additional explanations that encompass different dimensions. From the biographical perspective of the protesters, it was largely a movement with a significant presence of left-wing university students, even though the demands being promoted—better pensions, higher wages, improved public transportation, and better education—were broad in nature and resonated with much of the population. Despite this, it remained a somewhat disorganized movement, with sporadic leaders and civil society organizations—structural availability—playing a secondary role (Somma et al., 2021).

Nevertheless, one of the factors that makes this period particularly challenging to analyze is the wide range of demonstrations of discontent, spanning from peaceful protests to violent actions and common criminal activities, such as looting and the destruction of public and private property. As has been previously noted, the looting, which intensified by the end of October, even led to supply shortages, particularly affecting vulnerable communities.

What followed was an atypical political process in the country's history, marked by a broad political agreement that initiated a constituent process aimed at replacing the then-current constitution, which many in the political spectrum associated with Pinochet. In 2020, amid the COVID-19 pandemic, 78% of the population voted in a plebiscite to rewrite the constitution, with 79% opting for the task to be undertaken by a group of elected representatives. In this process, right-wing groups in Chile secured only a quarter of the seats. However, the narrative took an unexpected turn in September 2022, when 62% of the population decided to reject the proposed new constitution in a plebiscite.

As a result of this first failure, the political class reached a new agreement to push for a second constituent process, this time with 50 elected representatives working based on a proposal first agreed upon by a group of experts designated by political parties. In this second attempt, the right secured 33 of the 50 available seats. Nevertheless, in December 2023, 56% of the citizens again opted to reject the proposed constitution in a plebiscite. Consequently, Chile became the only country to reject two proposed new constitutions at the polls, with the political class failing to materialize an institutional solution to the crisis that has gripped the country since 2019.

1.3.2 Centralism, Marginalization, and Territorial Concentration

As of 2024, Chile is divided into 16 regions and 346 municipalities (see Fig. 1.1). Similar to other Latin American countries, Chile exhibits an extremely centralized structure, characterized by lagging regions, low subnational spending, and areas heavily

1.3 Case Study: Chile (2017–2023)

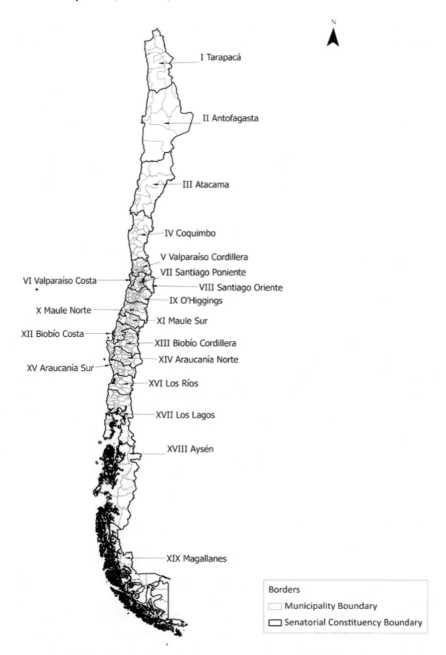

Fig. 1.1 Chilean map of Municipalities. *Note* This shows the continental municipalities of Chile, excluding two insular municipalities (Juan Fernández and Easter Island)

reliant on direct transfers from the capital (OECD, 2009, 2017). Additionally, the country experiences high economic inequality, which manifests both interpersonally and territorially.

This multifaceted inequality is crucial, as all analyses aimed at understanding the country's political crisis identify inequality as a principal issue. However, it is often overlooked that these inequities are not solely interpersonal but also territorial. In Chile, the latter issue is evident in the significant concentration of development benefits in Santiago. The capital city boasts the best schools, hospitals, clinics, universities, cultural events, and much more.

While territorial concentration is not a new challenge, it is alarming that it has been intensifying each year. This is perplexing, considering that since the dawn of our new democracy, there has been a consensus on the urgency and severity of the problem.

Reflecting on the social upheaval of October 18, 2019, the focus is often on the protests in Santiago (the capital). Rarely is it remembered that even earlier, President Sebastián Piñera's government had to contend with protests in Aysén, Punta Arenas, Freirina, Copiapó, and Chiloé, among others. These were all territorial conflicts rooted in a sense of abandonment. Some dimensions of the discontent perceived in Chile—those that Offe (2006) associates with more passionate elements—seem closely tied to this harsh feeling of being forgotten.

Therefore, if the predominant sentiment is one of abandonment, it is urgent to identify and understand the underlying elements of this "disaffection"—as termed by some political scientists. Recent research suggests that discontent—and consequently anti-establishment voting—can only be understood through the lens of territory, specifically by examining what happens in "places that don't matter". Given Chile's excessive centralism and territorial concentration, territorial inequalities take on a special significance in this discussion. Currently, there are inhabitants who feel marginalized, forgotten, and insignificant to the system. Unfortunately, the critical nature of this issue has not translated into public policies or genuine interest.

1.3.3 The Chilean Region of Valparaíso and its Territorial Challenges

As will be explained in subsequent chapters, the data used in this work pertains specifically to the Chilean region of Valparaíso, the second largest in the country. While this area can be seen as representative of the national reality, the region of Valparaíso also possesses unique characteristics. Notably, the city of Valparaíso (the regional capital) hosts the National Congress and was one of the focal points of the 2019 social uprising (BBC, 2019).

In the context of Chile's exacerbated centralism, certain territories also report suffering from intraregional centralism. This translates to typically smaller territories with low electoral weight being relegated to a secondary plane in discussions about regional development policies and measures. Specifically, authorities tend to focus on more central territories where there is a larger voter base. This reality is evident in various parts of Chile, and in recent years, some regions have even advocated for separation, accusing a regional elite—both social, political, and economic—that does not respond to local interests and cultures. Over the past decade, two new regions have been created. In the region of Valparaíso, this issue is particularly prominent, with a long-standing campaign aiming to create a new region in the current province of Aconcagua (currently part of the region of Valparaíso). These efforts were even brought to the recent constitutional process (BBCL, 2021; TVN, 2022).

In summary, Chile is a country characterized by both interpersonal and territorial inequality, exacerbated centralism, and currently undergoing one of the most significant political crises in its recent history. Concurrently, it is a nation with a certain level of development and high Internet penetration, which, as seen globally, has impacted the ways in which people engage politically.

1.4 Structure of the Book

To fulfill our purpose, this book is structured as follows. The next section will develop the theoretical framework of our proposal, starting with the challenge of discontent, followed by the specific issue of political disaffection, and then exploring studies that have addressed these topics through the role of digital platforms and territorial context. Following this, another section of the book will be dedicated to the methodological approach of this project. Within this framework, we will detail the development of the questionnaire, share specifics about the data used, and delve into the techniques employed to test our hypotheses and examine the described phenomena.

In a subsequent section, we will present some results based on a series of investigations conducted over the past five years. Finally, the concluding section will offer reflections on our findings and share broader insights.

References

BBC. (2019). Protestas en Chile — Valparaíso: Cómo la ciudad Patrimonio de la Humanidad se convirtió en un campo de batalla [newspaper]. Retrieved February 23, 2024, from https://www.bbc.com/mundo/noticias-america-latina-50643444

BBCL. (2021). ¿Región de Aconcagua? Subdere confirma que segunda etapa para evaluar independencia comenzará pronto [newspaper]. Retrieved August 15, 2023, from https://www.biobiochile.cl/noticias/nacional/region-de-valparaiso/2021/07/27/region-de-aconcagua-subdere-confirma-que-segunda-etapa-para-evaluar-independencia-comenzara-pronto.shtml

Castiglioni, R. (2014). Chile: Elecciones, conflictos e incertidumbre. *Revista de Ciencia Política, 34*(1), 79–104.

Castiglioni, R., & Kaltwasser, C. R. (2016). Challenges to political representation in contemporary Chile. *Journal of Politics in Latin America, 8*(3), 3–24. https://doi.org/10.1177/1866802X1600800301

CEPAL. (2016). *Estado de banda ancha en América Latina y el Caribe 2016*. United Nations.

Gamboa, R., & Segovia, C. (2016). Chile 2015: Falla política, desconfianza y reforma. *Revista de Ciencia Política, 36*(1), 123–144. https://doi.org/10.4067/S0718-090X2016000100006

Luna, J. P. (2016). Chile's crisis of representation. *Journal of Democracy, 27*(3), 129–138. https://doi.org/10.1353/jod.2016.0046

Mainwaring, S., & Scully, T. R. (2008). Latin America: Eight lessons for governance. *Journal of Democracy, 19*, 113–127. https://doi.org/10.1353/jod.0.0001

Morales Quiroga, M. (2021). Chile's perfect storm: Social upheaval, COVID-19 and the constitutional referendum. *Contemporary Social Science, 16*(5), 556–572. https://doi.org/10.1080/21582041.2021.1973677

OECD. (2009). *OECD territorial reviews*. Chile.

OECD. (2017). *Making decentralisation work in Chile*. https://www.oecd-ilibrary.org/content/publication/9789264279049-en

OECD. (2020). *Digital government in Chile—Improving public service design and delivery*.

Offe, C. (2006). Political disaffection as an outcome of institutional practices? Some post-Tocquevillean speculations. In M. Torcal & J. R. Montero (Eds.), *Political disaffection in contemporary democracies*. Routledge.

Roberts, K. M. (2015). *Changing course in Latin America: Party systems in the neoliberal era*. Cambridge University Press. https://doi.org/10.1017/CBO9780511842856

Segovia, C., & Gamboa, R. (2012). Chile: El año en que salimos a la calle. *Revista de Ciencia Política, 32*(1), 65–85.

Somma, N. M., Bargsted, M., Disi Pavlic, R., & Medel, R. M. (2021). No water in the oasis: The Chilean Spring of 2019–2020. *Social Movement Studies, 20*(4), 495–502. https://doi.org/10.1080/14742837.2020.1727737

Subtel. (2017). *Novena Encuesta de Acceso*. Subsecretaría de Telecomunicaciones: Usos y Usuarios de Internet.

TVN. (2022). Convención constitucional plantean consulta para crear Región de Aconcagua [newspaper]. Retrieved August 15, 2023, from https://www.24horas.cl/regiones/zona-centro/valparaiso/convencion-constitucional-plantean-consulta-para-crear-region-de-aconcagua

Valenzuela, S., Arriagada, A., & Scherman, A. (2014). Facebook, Twitter, and Youth engagement: A quasi-experimental study of social media use and protest behavior using propensity score matching. *International Journal of Communication, 8*. http://ijoc.org/index.php/ijoc/article/view/2022

Valenzuela, S., Arriagada, A., & Scherman, A. (2012). The social media basis of youth protest behavior: The case of Chile. *Journal of Communication, 62*(2), 299–314. https://doi.org/10.1111/j.1460-2466.2012.01635.x

Chapter 2
Theoretical Framework

Abstract This chapter explores the critical role of territorial contexts in shaping modern political sentiments of powerlessness and frustration. While discontent is often studied through individual-level factors, recent research highlights the profound influence of geographical and contextual elements. Building on concepts such as Katherine J. Cramer's "rural consciousness" and the "geography of discontent" framework, the chapter examines how territorial marginalization and historical economic trajectories influence political attitudes and behaviors, with case studies from Latin America, Europe, and the United States. It argues for a shift from the "geography of voting" to a "geography of discontent", underscoring the need to address attitudinal dimensions often overlooked in traditional analyses. The chapter delves into Latin American protests, such as Chile's 2019 demonstrations, to illustrate how unmet expectations and perceptions of neglect fuel political resentment. Unlike Europe and the U.S., where discontent often channels into populist narratives, the Latin American experience highlights diverse expressions of dissatisfaction, including protests and abstention. Furthermore, the chapter introduces tools for understanding the democratizing potential of digital platforms, particularly for marginalized groups. By integrating territorial and digital dimensions, this work provides a multidisciplinary framework to examine political disaffection, offering insights for addressing democratic challenges in diverse contexts.

Keywords Political disaffection · Political efficacy · Territorial inequalities · Digital divide · Civic engagement

2.1 Political Discontent: Specificity, Trajectory, and Discontent

2.1.1 *Political Discontent as a Multidisciplinary Phenomenon*

Despite the recent surge of interest from various disciplines in understanding and explaining political discontent, it is, in fact, a longstanding issue. Over recent decades, especially since the second half of the twentieth century, the study of political discontent has been a recurrent theme in social sciences. In both consolidated and emerging democracies, discussions about disaffected citizens (Torcal, 2006), a decline in civic engagement (Almond & Verba, 1989; Putnam, 1995), crises of representation (Luna, 2016), and abandoned places (Rodríguez-Pose, 2016) have become common. However, the most classical research on the subject has not always been accurate in predicting the potential consequences of these phenomena.

The impact of discontent on democracy has particularly intrigued numerous researchers. In their seminal work on democratic values and civic sense, Almond and Verba (1989) emphasized the importance of understanding political attitudes in conjunction with the effectiveness and stability of the democratic system in which they operate. They believed that both were intrinsically related and interacted with one another. Specifically, they argued that the functioning of institutions is influenced by various factors related to engagement, such as the frequency of civic or political participation and the belief that the system is legitimate and responsive to "ordinary citizens". This classical idea of engagement—which might range from Aristotle to Tocqueville—has evolved into a reasonable and consistent argument. Democracy would thrive on an active and participative citizenship in civic affairs, highly informed with a strong sense of responsibility. Not only is the system shaped by the beliefs and attitudes of its citizens, but these citizens should also act according to the "requirements" of the democratic system (Almond & Verba, 1989).

However, over time, it hasn't been entirely straightforward to test and demonstrate these ideas. As a result, some scholars have argued that the various problems associated with discontent haven't necessarily translated into any form of imbalance in democratic systems. This point is crucial, as various authors have suggested that studying political discontent implies embracing the complexity and multidimensionality of the phenomenon. Thus, depending on the approach taken, pinpointing its causes and consequences becomes ambiguous.

In their seminal work, "The Legitimacy Puzzle in Latin America", Booth and Seligson (2009) begin by challenging one of the classic assumptions about the study of discontent, namely that disaffection would have detrimental effects on the legitimacy of democratic regimes. Based on the information and data available to them at that time, they tentatively suggest that citizens who are displeased with the management of their governments and show dissatisfaction don't necessarily end up protesting or disengaging from traditional political activities. On the contrary, this very group—i.e., the disaffected—might be the segment of citizens that participates at higher rates, both through conventional mechanisms and alternative initiatives (such as civil society or community-based projects). In summary, what these authors convey is that, even when administrations are judged poorly, support for the democratic regime often remains robust and in good health.

Among many other things, Booth and Seligson's approach highlights a series of challenges faced when studying and understanding discontent. Firstly, as we've already hinted at in the preceding paragraphs, there's no denying that various approaches to the phenomenon differ both in theory and methodology. Many studies perceive discontent as a unidimensional phenomenon, linking it to more specific problems—such as dissatisfaction with a particular government, distrust in institutions, and frustration stemming from individual political action. Upon reviewing the significant contributions to date, it becomes clear that there are elements of discontent that don't necessarily follow the same dynamics and logic. For instance, there's a marked difference between a rational rejection of the democratic system and a more visceral feeling of anger and helplessness toward a specific authority or institution (Offe, 2006). Similarly, examining citizens' relationships with their local political communities isn't the same as analyzing their ties with national institutions (McKay, 2019). While all these aspects relate to what's typically associated with discontent, they differ in many ways. Their characteristics, evolutions, explanations, and outcomes vary, suggesting that they should be studied individually.

This problem of "specificity" is compounded by a second challenge related to the trajectory of the phenomenon under study. Often, there is a lack of quality data that allows for an analysis of political discontent, including its temporal dimension. This is crucial since the study of political attitudes should account for variations in their volatility. Classical authors, like David Easton, highlighted this point with great precision. While there might be beliefs and feelings closely tied to the performance of an administration, there are also attitudes that are more diffuse and stable, many of which persist over time, even intergenerationally (Easton & Dennis, 1967; Easton, 1975, 1976).

Beyond the challenges of "specificity" and "trajectory", there's still a gap concerning the true extent of the significant findings reported over the past decades. As one might imagine—and as Booth and Seligson (2009) suggest—discontent should be understood within its specific context when exploring its reasons and consequences. This underlines the importance of considering territories often absent from the debate—typically the Global South—as well as the need to delve into spatial aspects tied to local dynamics. Indeed, this difficulty is the primary motivation for this book. As we will detail in subsequent sections, recent studies have emphasized the importance of addressing discontent by incorporating specific territorial context, including intra-urban differences. It's not just about urban-rural or core-periphery divides, but also recognizing that political sentiments are experienced at the neighborhood level. Based on available evidence, living in a "neglected" region can influence certain attitudes, but this doesn't disregard the impact of smaller territorial units, not only in terms of belonging but also in neighborhood relations. Unfortunately, these topics haven't always been covered in the literature, mainly due to a lack of data that would facilitate studying sentiments and beliefs at the local level. Despite several global engagement projects, most available surveys—particularly relevant when measuring political attitudes—focus on national or continental populations.

However, these complexities inherent in studying discontent shouldn't lead to paralysis, quite the opposite. More than ever, we now have resources and techniques that allow us to face the challenges in this area.

In studying civic values, Almond and Verba focused on the rise of fascism and communism after World War I. In their view, this process would have raised genuine doubts about the inevitable future of European democracy. Their question was wholly pertinent at the time: Is it possible to identify processes suited to the unique cultures and social institutions of each country? According to the authors, World War II would have extrapolated this doubt globally, with new political actors pushing to break away from modern world isolation. This latter reflection remains interesting today. The authors don't just speak of countries or continents but of a broad group of historically marginalized individuals who, back then, were demanding entry into the political system.

Today, studies on discontent face a scenario that shares certain features with those described by Almond and Verba. The challenge of inclusion might be likened to what is experienced in various places today. Even within the most consolidated democracies globally, there's talk of forgotten places and people, of groups that have been left behind by progress.

When discussing discontent and unease, one must necessarily consider this facet of marginalization and neglect—commonly found in recent studies on the "Geography of Discontent"—which isn't necessarily tied to sociodemographic parameters. In the specific case of developing countries, this lag isn't only evident through a certain economic trajectory but also by various other factors such as territorial concentration, perceptions of inequity, gender, or even age. All these aspects—among many others—are crucial and essential when interpreting and explaining discontent.

2.1.2 Political Discontent in Latin America

And, once again, the relevance of studying these phenomena is faced with the challenges of accurately capturing them. In light of the triumph of populist, nationalist, and anti-European narratives, we've seen many efforts to understand the dynamics of discontent through electoral results. Specifically, the evidence suggests that the potential anger and helplessness felt by citizens have been harnessed by projects seen as a "threat" to the democratic system. As such, a series of liberal institutions that, in the vein of Almonf and Verba (1989), once seemed like accomplishments representing an "inevitable fact" are now being questioned by a range of political leaders with widespread popular support.

One of the main problems, however, is that these dynamics—particularly noticeable in various European countries and the United States—are not applicable to all global contexts. In some cases, for instance, discontent does not necessarily manifest through electoral processes. Before that, there's a wide variety of political actions that are characterized by deep penetration among citizens. Recent events in Brazil are proof of this. At the beginning of 2023, thousands of protesters stormed the Plaza of the Three Powers, located in the capital, Brasília. During the altercation, both the Supreme Federal Court and the National Congress were vandalized, with damages affecting numerous relics and challenging the very institutional framework as a whole. Just a couple of years earlier, in Colombia, hundreds of thousands of citizens took to the streets in protest. Defying the health measures in place during the pandemic—the events occurred in April 2021—the demonstrations were initially triggered by opposition to the tax reform project promoted by then-president Duque.

2.1 Political Discontent: Specificity, Trajectory, and Discontent

However, shortly after, the protesters clarified that their real motivations were somewhat more nebulous, attributed to policies implemented over the past decades in the South American country.

Chile, our case study, had experienced a process quite similar to that of Colombia just two years earlier, with significant consequences that persist up to the time this book is being written. On October 18, 2019, following a fare increase for the metro—the main urban transport in Santiago—the most massive and significant mobilizations since the return to democracy were triggered. In the country's capital alone, more than a million people gathered in the streets, with episodes of destruction, violence, and repression. As would later happen in Colombia, protesters stated that their actions were not solely due to the public transportation fare increase—an increase amounting to USD $0.035—but rather for structural reasons, embodied in policies they considered abusive and implemented over the last 30 years.

This reality, which seems very characteristic of recent Latin American history, is far removed from the way political attitudes—and, with them, discontent—are experienced in other parts of the world.

For some authors, regardless of the dispersion, extent, and frequency of protest activities, the social outbursts, which at times seem dramatic, would have a rather limited impact on the democratic system. According to Booth and Seligson (2009), for example, these would be episodes so sporadic and contained that they wouldn't even come close to weakening or destabilizing the regimes of the countries where they occur. In that regard, they don't question their occurrence, but they do, to some extent, question the magnitude of their consequences.

Given recent events, we are well aware that this idea seems to be, at least, controversial . What has been experienced in Brazil, Colombia, and Chile (and even in the U.S. and France, for instance) suggests that, under certain contexts, the protest activity associated with discontent could indeed have consequences that affect the stability of the regimes in the medium and long terms.

The case of Chile is perhaps the most evident among the three Latin American experiences. The protests of 2019, characterized by their scale and violence, led to an ongoing constituent process where the most fundamental rules of coexistence were discussed. Although various studies show that interest in the process has decreased considerably, the fact is that 4 years of constitutional discussion have already passed. This counterpoint is particularly interesting. Just as in the USA and Europe, in the case of Chile, certain populist narratives have also thrived, basing their projects on the discontent and unease experienced by large sectors. However, we have already witnessed that this discontent does not necessarily lead to an electoral solution. As some authors have suggested (Navarrete & Tricot, 2021), Latin American reality shows that discontent often manifests in diverse ways, which escape the conventional routes that attempt to channel it. For this reason, it is especially pertinent in these contexts to go beyond the vote, explaining and understanding attitudes and feelings that come before.

In conclusion, both classic and recent research on discontent shows us, at least, two aspects that can help create a good starting point in our reflection. First is that the phenomenon in question remains as relevant today as it did 50 years ago. Neither democracy nor liberal institutions seem to be "inevitable facts", largely due to the projects and narratives that are based on disaffection, neglect, anger, and powerlessness of part of the citizenry. And, second is that special attention must be paid to the methodological and theoretical challenges that the area poses. What we will work on next has to do precisely with that—a proposal that stands out for its specificity, considering the trajectory of the object studied, and providing new information that allows us to extend, even more, the scope of the important findings made to date. To the extent that we achieve that, the objective of this book will be fulfilled.

2.2 The Unique Problem of Political Disaffection

2.2.1 How to Frame Political Disaffection

Understanding that political malaise can be perceived as a phenomenon encompassing various dimensions and necessitating a degree of specificity, disaffection emerges as a significant facet.

Although many studies have delved into disaffection as a distinct dimension of unease, a clear definition of the phenomenon has not always been proposed. It is believed to be related to the process of distrust toward institutions and their leaders, leading to alienation from the political order (Maldonado Hernández, 2013). This perspective, deeply rooted in Giuseppe Di Palma's initial propositions (Di Palma, 1969, 1970), relies on the belief that democracy requires citizens to feel involved in political processes, as well as in the work of their institutions and representatives. Following this logic, disaffection—and the variables estimating it—have become indicators of the health of democracies, both as a component of civic virtue and as an underlying element of accountability.

While this book does not aim to reach a consensus on a definition of disaffection, it's necessary to define certain boundaries to distinguish this dimension from others similarly linked to discontent. In this vein, some authors—especially considering new democracies—have drawn a distinction between three aspects: legitimacy, discontent (or dissatisfaction), and disaffection (Montero et al., 1997, 2016). Following this, sociologist Claus Offe provides an insightful perspective that helps us discern differences among them:

> If my interests are being violated, I am left with a sense of dissatisfaction; if the reasons given for the worthiness of the political order and its actual governance practices aren't supported and confirmed by autonomous insight, we speak of illegitimacy, experienced as a lack of valid reasons supporting what we observe in public policies and their impact on 'us'; and if people detach themselves from a polity or political community they perceive as unfamiliar, boring, incomprehensible, hostile, or unreachable, we can term it as disaffection. (Offe, 2006, p. 25)

Thus, disaffection would be closely associated with a more passionate element, a feeling of dissociation from a system perceived as inaccessible and unresponsive to citizens' demands and interests.

2.2 The Unique Problem of Political Disaffection

In this regard, and as previously stated, it seems that the dimension of political malaise will depend on the specific political object in question. Thus, it could be argued that disaffection pertains to the relationship between the citizen and institutions. Therefore, we're not necessarily talking about a rational rejection of the democratic system—a factor more tied to legitimacy—nor an evaluation of the performance of a particular authority—a factor more tied to dissatisfaction—but rather about citizens distancing themselves from political institutions (Maldonado Hernández, 2013). This idea is crucial for understanding the phenomenon of political malaise in new democracies. For several years now, many citizens appear to be redefining their relationships with the political process, its institutions, and the political elites who govern, but this doesn't necessarily translate into questioning the democratic order (Torcal & Montero, 2006a).

Given the characteristics of the phenomenon, it might be suggested that, unlike dissatisfaction, disaffection could be considered more stable over time, without entirely depending on the variations in the policies implemented by a particular government or the popularity level of any given authority.

As we'll notice, many of the phenomena discussed in previous sections—such as the resurgence of populisms, nationalist views, and anti-European sentiments—would be more related to feelings associated with disaffection, rather than dissatisfaction or perceptions of legitimacy. Though we'll revisit this idea later, as early as 2006, Claus Offe warned that, in response to these beliefs, the political class tended to try to bridge this gap with rather demagogic appeals to certain cultural values and the emotions attached to them—like indignation. In this sense, a pattern he found familiar was the emergence of politicians acting as "anti-political politicians" (Offe, 2006).

However, as we have previously suggested, the explanations and consequences of these processes must be understood within their own context. Perhaps for this reason, the level to which these feelings and beliefs are undesirable has always been questioned. While at times they are associated with a reluctance to inform oneself or with low participation (Montero et al., 1997; Torcal & Lago, 2006), other authors suggest that the existence of critical citizens is essential in the search for new relationships and greater accountability (Nye et al., 1997). In other words, and at the risk of oversimplifying a more complex phenomenon, we could understand that, on the one hand, mistrust or cynicism regarding the "political class" could be seen as a syndrome that contemporary democracies have been suffering from for some time, but when directed correctly, it could even strengthen them, prompting the search for additional alternative modes allowing for greater mobilization and representation. On the other hand, the same author adds that unchecked political disaffection has the potential to create spaces and opportunities that could be exploited by populist projects, with their consequent illiberal and antidemocratic narratives.

So, although we have chosen not to offer a specific definition of disaffection, we have shared certain frameworks that help us identify some of its characteristics—i.e., mostly emotional and stable feelings and beliefs that result in a dissociation between citizens and a political system (with its institutions and authorities) perceived as unresponsive. Following this line of reasoning, some authors suggest that the phenomenon would consider at least two aspects with potentially different consequences. One is linked to a lack of involvement and generalized distrust regarding politics (i.e., political disengagement). The other is associated with a lack of reciprocity on the part of the system (i.e., institutional disaffection). Both refer to different things and therefore would be measured differently. Thus, their explanations and consequences would also be specific.

All of this discussion helps provide specificity to the study of political malaise. If by "discontent" we refer to the anger, powerlessness, or frustration experienced by certain people—in certain places—who feel abandoned by the system, then it's likely we're not talking about dissatisfaction (discontent) or illegitimacy, but rather about disaffection. Understanding these differences is crucial. In Latin America and in some of the new European democracies, it has been shown that unconventional forms of political participation do not represent a response to the dissatisfaction generated by conventional politicians, but rather the result of decades of upheaval experienced by generations of citizens in the past. That would be disaffection (Torcal & Lago, 2006).

In what follows, we will offer concrete alternatives to capture the feelings associated with both dimensions of disaffection, that is, those linked to political disengagement and institutional disaffection.

2.2.2 The Personal Belief of Self-competence

As we have pointed out, disaffection could be understood as one of the specific dimensions of political malaise. Following this line, disaffection would be associated with specific feelings and beliefs that can be more accurately captured and measured.

Under Torcal and Montero's view, both political interest and trust would be, for example, two political attitudes associated with the particular phenomenon of disaffection. The former would help us capture what the authors refer to as political disengagement and the latter what they call institutional disaffection (Torcal & Montero, 2006b).

However, it is also suggested that other attitudes could complement this analysis. Specifically, it is often recommended to use political efficacy to capture other types of feelings that are equally structural and passionate.

Political efficacy is an attitude measured since 1952 by the Center for Political Studies (CPS) at the University of Michigan. The definition of the concept is often attributed to Campbell, Gurin, and Miller. In their emblematic work, "The voter decides", the authors define it as "the feeling that individual political action does have, or can have, an impact upon the political process, i.e., that it is worthwhile to perform one's civic duties" (Campbell et al., 1954, p. 187). The formula they suggested to capture these "feelings" was specific. They proposed a set of five statements that would, in turn, form an indicator for the attitude. Thus, respondents had to answer how much they agreed or disagreed with the following items:

2.2 The Unique Problem of Political Disaffection

1. I don't think public officials care much about what people like me think.
2. The way people vote is the main thing that decides how things are run in this country.
3. Voting is the only way that people like me can have any say about how the government runs things.
4. People like me don't have any say about what the government does.
5. Sometimes politics and government seem so complicated that a person like me can't really understand what's going on.

With the answers to these questions, the authors identify those who feel greater efficacy or, alternatively, greater political futility. Thus, one of the first analyses they share relates to the positive interaction between political efficacy and participation, which they suggest is "dramatically" demonstrated. Also, from a more descriptive approach, Campbell et al. report the differences in political efficacy according to different sociodemographic segments. Both education and sociodemographic status (including occupation and salary) would be highly related to the studied political attitude, extending to gender and race.

The conclusions of their preliminary analyses are presented as follows:

> In summary, it is reasonable to conclude that citizens who feel that public officials are responsive and responsible to the electorate, who think that individual political activity is worthwhile and capable of influencing public policy, and who see that the private citizen's channels of access to government decision-makers are not confined to the ballot box, are much more likely to be politically active than those citizens who feel largely overwhelmed by the political process (Campbell et al., 1954, p. 194).

Up to this point, efficacy was treated as a unidimensional attitude that served to capture that feeling that political action is "worthwhile". However, a few years after Campbell et al.'s seminal studies, researchers used the concept but with certain additional specifications. For Coverse (1972), for example, political efficacy did not represent a monolithic attitude. For this reason, it was necessary to make a distinction between, at least, two closely related but different aspects. He called one "personal feelings of political competence" and the other "trust in system responsiveness". Something similar was proposed by Lane, who believed that the then-new concept of political efficacy clearly had two components, "the image of self and the image of democratic government" (Lane, 1959).

It is precisely this division that largely persists to this day.

Coverse's work was essential for what was later developed in the discipline. Just a couple of years later, Balch (1974) used his proposal to coin the classic concepts of internal political efficacy and external political efficacy.

By internal political efficacy, the author specifically referred to those individual sensations of political self-competence already described by Coverse, which he theoretically associates with individual ego strength. The scope of this political attitude was significant.

So far, Almond and Verba (1989), in their work originally published in 1963, characterized that group of citizens—i.e., participants—who were presumed to be attentive and informed about the political system as a whole, both in its political and governmental aspects. Exploring the different political cultures present in certain territories, the authors characterized a type of citizen who believes in himself as a "capable" actor to participate, contrasting with others who do not hold that same belief. They called them the "self-confident citizen". In the authors' own words:

> The self-confident citizen appears to be the democratic citizen. Not only does he think he can participate, he thinks the others ought to participate as well. Furthermore, he does not merely think he can take a part in politics: he is likely to be more active. And, perhaps most significant of all, the self-confident citizen is also likely to be the more satisfied and loyal citizen (Almond & Verba, 1989, p. 207).

The proposal to deconstruct the sense of political efficacy allows for a slightly more precise contrast of these ideas.

For Balch (1974), internal efficacy can be defined as an individual's belief that means of influence are available to him. Subsequent research also understands it as the sense of one's competence to understand and participate in public affairs (Craig et al., 1990; Craig & Maggiotto, 1982; Miller et al., 1980). Regardless, what is interesting is that it is an attitude measuring a personal feeling, not necessarily an objective parameter of political knowledge. When studying political malaise, this becomes particularly interesting, as it allows us to incorporate individual expectations into the analysis, which, to varying degrees, could be triggers of frustration and resentment.

The ways in which this sense of self-competence has been measured have varied over time. Even today, there are specific approaches and, also, some imprecisions. Morrell (2003) provides a detailed exposition of its evolution since the concept's inception. Based on his tracking, perhaps one of the most interesting milestones occurs in the 1987 pilot study of the National Electoral Survey (NES), which tests a series of new items that could capture the previously described feeling.[1]

After this testing, and considering the final items included in the 1988 version of the NES, Niemi et al. (1991) conclude that an appropriate way to measure internal efficacy could consider the following new 4 items:

1. I consider myself to be well-qualified to participate in politics (SELFQUAL)
2. I feel that I have a pretty good understanding of the important political issues facing our country (UNDERSTAND)
3. I feel that I could do as good a job in public office as most other people (PUBOFF)
4. I think that I am better informed about politics and government than most people (INFORMED).

---

[1] The following items were considered: (1) I consider myself well-qualified to participate in politics, (2) I feel that I have a pretty good understanding of the important political issues facing our country, (3) Other people seem to have an easier time understanding complicated issues that I do, (4) I feel I could do as good a job in public office as most other people, (5) I often don't feel sure of myself when talking with other people about politics and government, and (6) I think that I am as well informed about politics and government as most people.

2.2 The Unique Problem of Political Disaffection

Despite this approach, subsequent studies (Valentino et al., 2009) have also used the single item that belonged to the original questions:

1. Sometimes politics and government seem so complicated that a person like me can't really understand what's going on (COMPLEX)

With more or fewer similarities, and adapting to the contexts in which they are measured, all these questions have indeed been a cornerstone for studies to date. However, that doesn't mean there aren't some less conventional proposals.

For instance, some authors simply use the term "efficacy" to refer to this individual sensation of competence (Chan et al., 2016; Diehl et al., 2016; Stoycheff et al., 2016; Zhang et al., 2010). Others, on the other hand, use a different conceptualization for similar matters—like self-efficacy (Chen et al., 2019).

Likely due to the lack of surveys specially designed to measure this attitude—a problem highlighted by authors like Finkel (1987)—a recurring issue is the use of only a single item (which often differs across studies) to capture these beliefs.

As we can already discern, internal efficacy has become a widely used variable across different disciplines, being essential when measuring civic values. Many contributions have been made on the topic to date. Initially, some authors warned that sociodemographic elements—such as educational level and race—were related to the development of this attitude (Abramson, 1983; Acock & Clarke, 1990). Valentino et al. (2009), for instance, suggest that internal efficacy drives participation, in part, by facilitating anger. Finkel (1985; 1987), years earlier, highlighted something then novel. According to his results, he suggests that the relationship between internal efficacy and political participation is rather reciprocal. It's not just an attitude that explains and impacts various types of participation, but these types of participation also reinforce the individual sensation of citizens' competence. In particular, the author emphasizes the effect of participating in political campaigns, which could strengthen levels of internal efficacy. Hansen and Tue Pedersen (2014) report something similar, underscoring the democratic role of campaign activities. Analyzing the Danish parliament campaigns of 2011, they suggest that exposure to electoral activities would not only make citizens more interested and knowledgeable but also more efficacious.

Regarding the use of social media, researchers suggest, for example, that political participation primarily concentrates on those who already feel confident about their political knowledge. All these contributions, among many others, have helped us better understand that certain sensations and attitudes deserve to be studied specifically, with theoretical and methodological rigor.

2.2.3 The Personal Belief of System Responsiveness

As we've already suggested, ever since Lane's early works in the late 1950s, it's been understood that the individual sensation of competence would be just one of the dimensions making up the concept of political efficacy. If this dimension is related to the individual, there would also be a dimension related to the outside, to the democratic government. This is what Coverse (1972) named "trust in system responsiveness" and what Balch (1974) called external efficacy.

As other authors also point out, external efficacy is tied to the perception that the system and its institutions are responsive to citizens' demands (Craig et al., 1990). The focus, then, shifts from how competent the citizen feels to how permeable the system is.

Just as with internal efficacy, the sensation of system responsiveness has also been measured in various ways. In this regard, studies motivated by the National Election Study (NES) have been crucial in identifying a proper way to estimate this attitude. According to Niemi et al. (1991)—using the 1988 version of the NES—an appropriate factor for external efficacy could be constructed using the following items:[2]

1. People like me don't have any say about what the government does (NOSAY)
2. I don't think public officials care much what people like me think (NOCARE).

To these two questions originally included in the 4 items from the Center for Political Studies (CPS) at the University of Michigan—and also considered in Balch's original work (1974)—the authors added two more related to responsiveness:

3. How much do you think that having elections makes the government pay attention to what the people think? (ELECRESP)
4. Over the years, how much attention do you feel the government pays to what people think when it decides what to do? (GOVRESP)

Some authors have reported that, generally speaking, the indicator for external efficacy tends to be less stable than that of internal efficacy (Acock & Clarke, 1990). A contentious aspect when measuring external efficacy is distinguishing this attitude from mistrust toward institutions (Shingles, 1981). Craig and Maggiotto (1982) highlighted this nuance. They posited that citizens are more likely to trust a democratic government that responds to their demands, thus implying that external efficacy and trust might be intrinsically linked. However, Craig et al. (1990) argued that external efficacy is distinct from political trust. They suggest that efficacy should be measured in terms of the fairness of political procedures and outcomes, rather than how elites respond to citizens' demands.

This matter is significant, and its effects can still be seen in measurement decisions. For instance, according to Niemi et al. (1991), the following statement would serve to construct a trust factor:

> Do you think that people in the government waste a lot of money we pay in taxes, waste some of it, or don't waste very much of it?

---

[2] Other authors suggest a different approach. For instance, Shingles (1981) distinguishes between external efficacy concerning incumbents and external efficacy concerning the regime. This approach was tested in the 1987 pilot version of the NES. Craig et al. (1990) suggest that if there is a difference between the two, it is rather subtle.

2.2 The Unique Problem of Political Disaffection

However, other researchers Hansen and Tue Pedersen (2014), drawing on classic definitions of external efficacy, suggest that the following question is suitable for capturing system-responsiveness:

The politicians waste a lot of taxpayers' money.

In this context, it seems even more crucial to consider the unique circumstances in which research is conducted. In some cases, the distinction between one attitude and another might take on special nuances.

Nevertheless, despite the importance of the concept, interest in explaining and understanding external efficacy hasn't reached the magnitude that internal efficacy has achieved. In many contexts—especially when analyzing digital democracy—external efficacy has been less explored.

Partly, this could be due to the challenges in correctly measuring feelings of system responsiveness. Historically, it has been associated with distrust variables, and in recent years, with variables related to "populist attitudes". However, both classic and recent literature have made it clear that they are distinct phenomena, each with their possible explanations and potential consequences (Craig et al., 1990; Geurkink et al., 2019).

Regardless, external efficacy has become a valuable variable when studying political cynicism and elite-challenging behaviors (Craig, 1980), as it precisely captures the lack of reciprocity in systems.

2.2.4 Back to the Classics: The Gamson's Hypothesis

As we've discussed, for several decades it is been understood that variables related to political efficacy are among the most relevant in determining the health status of democratic systems (Craig et al., 1990; Niemi et al., 1991).

In a classic work by William A. Gamson, the researcher asserted that a "combination of high sense of political efficacy and low political trust is the optimum combination for mobilization–a belief that influence is both possible and necessary" (Gamson, 1968, p. 48). This idea, named the Gamson's Hypothesis, has been repeatedly tested in various contexts. Even in recent years, efforts have been made to contrast this thesis in its original terms, that is, using efficacy and trust. However, it's evident that the initial idea proposed by Gamson presents challenges, not only in understanding efficacy as a unidimensional concept—associated with the idea of possible and necessary influence—but also concerning the vagueness of trust, which can be interpreted in different ways. Therefore, over time, there's been a reinterpretation of Gamson's original hypothesis, seeking to clarify certain concepts inherent to the original spirit of the proposal. In this regard, Craig (1980) argues that, instead of debating about efficacy and trust, it would be more valuable to examine the gaps that might exist between those with high internal efficacy and low external efficacy. That is, between citizens who perceive themselves as competent to participate in politics—regardless of whether they really are or not—but who believe that the political system is impermeable and disregards citizens' interests. According to Craig, this scenario would be optimal for seeking alternative mechanisms to participation.

This reinterpretation proposal has been tested in various contexts, including Chile, our case study. According to Torcal and Montero (2006b), unlike other new democracies, the reinterpreted Gamson hypothesis (i.e., understanding it as the gap between internal and external efficacy) does make sense in Chile, negatively impacting the propensity to participate in conventional mechanisms. However, the authors' results did not suggest a positive relationship between the gap and unconventional participation.

Regardless, revisiting Gamson's ideas remains a challenge since they allow for the exploration of something akin to political frustration. Here, the concept of "trajectory" becomes meaningful, as a sustained gap over time could precisely represent sustained political frustration. We refer to the accumulation of such feelings as political resentment. This perspective is particularly compelling as it departs from a view of frustration and resentment linked to absolute deprivation. In other words, it's not just material shortcomings that would explain these negative sentiments, but also expectations and self-perceptions.

In our view, through traditional political efficacies (internal and external), we could attempt to understand various phenomena occurring in both new and consolidated democracies. Much is said about different emotions, but conceptual and methodological frameworks to estimate them are not always provided.

2.3 Digital Platforms and Their Democratic Role

2.3.1 Internet's Emergence: First Glances at Its Democratic Impact

Although political efficacy has long been viewed as a crucial attitude to determine the health status of democracies, nowadays, it becomes challenging to approach it without considering the digital realm. The opportunities and threats provided by digital tools—no longer so new—are well-acknowledged. While governments are making significant strides toward modernization and digitalization—for Chile, see, for instance, OECD (2020)—we are well aware of risks deemed to be the most pertinent of our time. For example, a recent report by Reporters Without Borders identifies disinformation as the foremost risk to contemporary democracies. Lately, with the advancements in artificial intelligence techniques, there is talk of a burgeoning "industry of simulacra". These phenomena relate to our previous discussions. In response to citizens feeling powerless in the global political context, it is suggested that the Internet might encourage the rise and expansion of populist movements aiming to "return power to the people" (Boulianne et al., 2020).

However, apprehensions about the civic consequences of the "electronic revolution", as termed by Robert Putnam, have existed for several decades. For instance,

2.3 Digital Platforms and Their Democratic Role

Putnam used the telephone industry as an example to illustrate his argument. Despite promises of overcoming physical barriers and rethinking what we understand by "neighborhood", the reality is that it merely reinforced those bonds made outside of that technology's use. "One does not meet new friends on the telephone", the author asserted. The dichotomy and perceptions regarding the Internet are aptly captured in the following passage:

> One central question, of course, is whether "virtual social capital" is itself a contradiction in terms. There is no easy answer. The early, deeply flawed conjectures about the social implications of the telephone warn us that our own equally early conjectures about the internet are likely to be similarly flawed. Very few things can yet be said with any confidence about the connection between social capital and Internet technology. One truism, however, is this: The timing of the Internet explosion means that it cannot possibly be causally linked to the crumbling of social connectedness described in previous chapters. Voting, giving, trusting, meeting, visiting, and so on had all begun to decline while Bill Gates was still in grade school. By the time that the Internet reached 10% of American adults in 1996, the nationwide decline in social connectedness and civic engagement had been under way for at least a quarter of a century. Whatever the future implications of the Internet, social intercourse over the last several decades of the twentieth century was not simply displaced from physical space to cyberspace. The Internet may be part of the solution to our civic problem, or it may exacerbate it, but the cyberrevolution was not the cause. (Putnam, 2000, p. 170)

So, for Putnam, the matter was not simply a debate between optimistic or pessimistic views, but rather about moving beyond utopian or grievance-based perspectives. The challenges facing democracies wouldn't be explained by the advancement of technological tools, but by deeper elements. However, this does not mean that we cannot study the role digital tools are playing in either exacerbating or mitigating these foundational issues.

In this context, many authors have explored what is commonly termed the "digital divide", which refers to the gaps between those who benefit from new technologies and those who don't (Wilson et al., 2003). In essence, the unequal access to digital tools means that information flows in higher socioeconomic segments grow significantly compared to the lower ones. Yet, it is not just a matter of material resources. Various factors contribute to these gaps, including family structure, generational differences, disabilities, race, and, of course, geographical location (Bonfadelli, 2002; Di Maggio et al., 2004; Friemel, 2014; Graham et al., 2018).

Today, the scenario has changed. Digital tools have become widespread, and while there are still countries with minimal Internet penetration, many regions have overcome the initial challenges of material access. However, widespread adoption of digital tools doesn't necessarily overcome the digital divide. As some authors suggest, mere information availability doesn't guarantee its effective use (Scheufele & Nisbet, 2002). Hence, studying the digital divide goes beyond just a "first-level" of access; it also includes examining the nature of use, user skills, and attitudes toward technology. Again, it's been shown that many sociodemographic factors are linked to the uses assigned to platforms; some of these enable the expansion of individual capital—in various forms—while others do not (Bonfadelli, 2002; Hargittai & Hinnant, 2008; van Deursen & van Dijk, 2014).

Over two decades ago, Norris (2001) meticulously outlined the multi-faceted nature of the "digital divide". At its core, this divide speaks to the differing levels of access and engagement individuals have with digital technologies. Norris broke this down into three distinct facets:

- **The Global Divide**: This refers to the stark discrepancies in Internet access when comparing industrialized nations to their developing counterparts. In simpler terms, while the former may have high-speed broadband as a standard, the latter might still be grappling with fundamental connectivity issues.
- **The Social Divide**: Even within a single nation, there can be pronounced gaps in digital access. This internal rift is often driven by socioeconomic factors, where the affluent have greater digital access and fluency compared to the economically disadvantaged.
- **The Democratic Divide**: Beyond just access, this facet delves into how people use their digital resources. Are they passive consumers or active participants in public life? Norris' work emphasized the distinction between those who employ digital tools for civic engagement and those who remain on the sidelines.

Building on Norris' foundational work, subsequent scholars have delved deeper, uncovering a myriad of sociodemographic factors that influence the digital divide. These factors range from one's education and age to gender and overall socioeconomic status (Best & Krueger, 2005; Gibson, 2009; Min, 2010). Recent scholarship suggests that beyond these surface factors, deeper psychological and cognitive elements play a role. For instance, an individual's internal sense of efficacy—believing that they can make a difference—or their views on digital privacy can impact their online behavior (Chan et al., 2018; Hoffmann & Lutz, 2019).

Another nuance comes from the type of Internet access. In a study conducted in Chile, a distinction was observed based on the devices used. Those accessing the Internet via mobile phones, predominantly from less affluent sections, largely engaged in recreational online activities. In contrast, individuals accessing the Internet from homes or offices—typically from wealthier segments—tended to seek informational or political content (Fierro et al., 2020).

The digital divide, particularly the "democratic divide", has fueled a lively academic discourse around the role of the Internet in democracies. Some posit a more optimistic "mobilization thesis", suggesting that the digital realm offers a unique opportunity to engage typically marginalized communities (Carrara, 2012; Hirzalla et al., 2011). They argue that the Internet has the potential to kindle civic interest even among the politically apathetic or those economically disenfranchised (Espinal & Zhao, 2015; Morris & Morris, 2013).

Yet, a counter-narrative exists. Some scholars caution against an overly rosy view, putting forth the "reinforcement thesis". This perspective warns that the Internet might simply strengthen the voice of those already engaged, leaving the marginalized further behind. Instead of being a democratizing force, the Internet, they argue, risks becoming just another tool in the arsenal of the already powerful, exacerbating existing societal inequities (Kneuer & Datts, 2020; Schlozman et al., 2010).

2.3.2 Political Efficacy and Digital Platforms

In the realm of political communication, scholars frequently examine citizens' attitudes, feelings, and beliefs. For instance, a study by Hansen and Tue Pedersen (2014) reveals that campaign actions are associated not only with increased trust, knowledge, and interest but also with a boost in feelings of self-competence and perceptions of system responsiveness. In essence, the authors argue that electoral campaigns don't merely influence voting decisions and directions; they also bolster the health of our democratic system.

This particular study underscores the profound influence communicative actions can have on shaping attitudes that are typically stable over time. Given this, when evaluating the democratic role of digital tools, it's unsurprising that political efficacy often emerges as a critical focal point. After all, these feelings of efficacy are intrinsically tied to both offline and online participation (Oser et al., 2022). In a comprehensive review of 193 prior articles (most of which only tangentially addressed the topic), Boulianne et al. (2023) postulate the existence of a virtuous circle. This circle links both forms of political efficacy to online political participation. They suggest that information gleaned from the Internet fosters feelings of being well-informed, which, in turn, fuels greater online participation. However, they also caution that this dynamic might shift based on various contexts.

The perspective offered by these authors is illuminating, particularly as it sheds light on the global sentiments of powerlessness that many feel. Yet, the subject matter gains added layers of complexity when one considers the challenges frequently encountered in such research. Beyond the already-acknowledged intricacy of measuring political efficacies—where the authors themselves emphasize inconsistent approaches to crafting sets of questions for different variables—the "digital divide" studies caution us that not all types of participation are created equal. In simple terms, using networks for entertainment yields very different outcomes than using them for information or public engagement. We delved into this particular issue in a study aimed at discerning the varied effects of participation on both internal and external political efficacy. The data from that research affirmed that there indeed is a distinct difference between civic and non-civic use (the latter term referring to entertainment or social use). While civic use correlated with a greater inclination to feel competent and believe in the political system's reciprocity, we found no significant results for social or entertainment use (Fierro et al., 2023).

In this vein, the impact of informational use on such attitudes has been well-documented. In a rather pioneering work, Wagner et al. (2017) demonstrated that digital information consumption not only influences higher levels of external efficacy but also posited that this is contingent on the specific institutional context. In essence, citizens who source their information from digital platforms tend to believe that the system responds to them. This correlation strengthens in more democratic contexts. Such findings align with what we previously asserted: when examining political attitudes, it's imperative to contextualize them.

To circle back to the core issue, it becomes somewhat questionable to assume that digital tools, regardless of their specific use, can positively impact political efficacies and, consequently, disaffection. Hence, efforts to understand these relationships necessitate a higher degree of specificity—both theoretically and methodologically—and a broader contextual lens.

2.3.3 The Concept of Online Political Efficacy

Arising from studies on digital media and political efficacies, there emerged the idea that, through the Internet, citizens might feel closer to authorities and institutions. This closeness potentially enhances interactions and, as a result, the accountability processes (Kenski & Stroud, 2006). However, we have previously mentioned that researching the relationship between the Internet and political efficacy has confronted various challenges. The most frequently cited challenge pertains to difficulties in agreeing on how to pose and estimate questions about political attitudes. Yet, as some authors suggest, there's an additional problem: the failure to consider the unique context of the digital realm. While certain researchers assert that the effects of efficacies remain consistent across online and offline worlds (Oser et al., 2022), others underscore the need to study political efficacy in both the online and offline domains distinctly. In essence, they propose that when gauging efficacy, responses might differ depending on whether the respondent factors in digital tools when answering (Sasaki, 2016).

Given this context, the past decade has seen the introduction of an efficacy measure specifically tailored for the online world, termed "Online Political Efficacy". The underlying thesis is straightforward: if the questionnaire embeds the phrase "by using the Internet" when framing the response, such answers should be more accurate in predicting online political participation.

Thus, the World Internet Project has incorporated a set of questions specifically designed to capture this new dimension. Specifically, they employ a 5-point Likert scale in response to four questions:

1. By using the Internet, people like you can have more political power.
2. By using the Internet, people like you can have more say in what the government does.
3. By using the Internet, people like you can better understand politics.
4. By using the Internet, pubic officials will care more about what people like me think.

While a topic we will discuss in greater depth later, it is crucial to note that the questions posed are not necessarily tailored to distinguish between the traditional concepts of internal and external efficacy. Instead, they seek to gauge the sentiment that political action is worthwhile in the digital environment, aligning more with a broader notion of general efficacy. Despite not being a universally shared viewpoint,[3] it seems that the central and distinguishing element of these questions pertains to the use of the Internet.

Sasaki (2016) demonstrates that these questions yield a novel and consistent index, both internally and externally, which can be justified on theoretical and empirical grounds.

To date, suggestions have been made that this new measure of efficacy could play a significant role in mediating the relationship between the use of social media and online political participation (Chen et al., 2019). Additionally, it has been shown that other beliefs—such as the credibility of the press, tolerance toward press freedom, and the trustworthiness of online information—are positively correlated with online efficacy levels (Martin et al., 2018).

Moreover, this new efficacy metric has offered a fresh perspective on studying the Internet's role in integrating sectors typically marginalized from public affairs. Sasaki (2017) indicates that in lower education sectors, the utilization of digital tools correlates more strongly with higher online efficacy levels than in high-education groups. Similarly, Fierro et al. (2023) have shown that those residing in politically peripheral areas—those geographically distant from political decision-making centers—are more inclined to feel that, due to the Internet, the system becomes more approachable and receptive.

This latter notion will be explored in more depth when we present our analysis results and, in some respects, epitomizes one of the central themes of this work. Thus far, we could postulate that feelings of powerlessness and alienation from the political system are critical when examining contemporary democratic challenges, especially in explaining the rise of populist, anti-establishment, and nationalist (or anti-European) narratives. We have also proposed that political efficacies might provide insights into these feelings of neglect. The arguments presented in this chapter are intertwined with the role of the Internet in these relationships.

2.4 Territory and Political Engagement

2.4.1 Political Engagement and Space: Some Initial Considerations

Recent years have observed a rise in political disaffection, characterized by feelings of powerlessness and rage in the political realm. However, this discontent is not solely the result of individual sociodemographic factors. The academic discipline of political science suggests that phenomena such as corruption, political scandals, and the performance of institutions can influence the emergence of these disappointed

---

[3] For a more detailed discussion on this point, it is suggested to consult the initial studies by Sasaki (2016; 2017).

expectations. In contexts characterized by centralized power structures, as is the focus of our study, many elements of this discontent can be explained from a territorial perspective.

Over five decades ago, Lipset and Rokkan (1967) introduced the idea of social cleavages to decipher the factors influencing the establishment and stabilization of party systems in mature democracies. A significant concept within this framework is the center-periphery cleavage, which highlights the cultural, religious, economic, and historical distinctions between the central forces that constitute nations and their provincial counter-movements. This cleavage accentuates the disparities between those at the epicenter of social, political, and economic power and those on its fringes. Subsequent academic work has explored this further, particularly in the context of interregional relationships and international relations theories (Cardoso & Faletto, 1976; Wallerstein, 1974).

However, when examining political involvement and participation, the focus has often shifted from the center-periphery cleavage to the urban-rural divide and the role of territory size (Andrews et al., 2019; McDonnell, 2020; Rodrigues & Meza, 2018; Saglie & Vabo, 2009). Historically, the topic of the optimal size of territorial units, commonly interpreted by population size, has fascinated scholars since the dawn of democracy. With the global urbanization surge in the mid-twentieth century, early warnings emerged suggesting that the size, diversity, and density of territories might erode the interpersonal and psychological bonds among citizens (Wirth, 1938). On this note, Dahl (1967) proposed that smaller regions might foster greater civic participation in governmental decisions. In his view, as the territorial unit expands, it becomes less manageable, leading citizens to lose their sense of community, fostering feelings of detachment and inefficacy. Yet, Dahl also contended that in exceedingly small settings, the influence exerted by citizens becomes inconsequential, which might hamper participation (Dahl, 1967; Dahl & Tufte, 1973). The matter grows intricate with some scholars reporting higher local political participation in smaller territories, yet this might not translate to national involvement (Nie et al., 1969). Fischer (1975) further elaborates that the size of the administrative unit might impact participation in diverse ways. Larger regions might experience heightened political inefficacy but simultaneously greater mobilization on national issues (Fischer, 1975). Considering that larger territories attain higher efficiency levels, some literature indicates an overemphasis on the merits of smaller units, fueled by the romanticized belief that "small is beautiful" (Newton, 1982). While the debate remains contentious, evidence from Oliver (Oliver, 2000) supports Dahl's early viewpoint, suggesting that larger American cities experience lower propensities for civic engagements such as contacting authorities or participating in local elections. Recent studies from Portugal echoed these sentiments, indicating that territorial integration processes negatively affected electoral participation (Rodrigues & Meza, 2018).

2.4.2 The Role of Territory on Civic Participation Patterns

Undoubtedly, territory plays a pivotal role when examining patterns of civic participation, whether traditional or digital. In recent years, especially against the backdrop of rising discontent and the ascent of populist narratives in Europe and North America, there have been notable contributions centered around the "Geography of Discontent". In this context, spatial considerations are deemed essential for understanding this phenomenon. Anti-establishment sentiments predominantly emerge from what have been termed "places that don't matter" (Rodríguez-Pose, 2016) or "people and places left behind" (Becker et al., 2017; Goodwin & Heath, 2016). These sentiments do not necessarily originate from impoverished or affluent areas, but rather from regions that, once prosperous, now grapple with feelings of abandonment (Rodríguez-Pose, 2020). Against this backdrop, territory assumes a central position when probing the genesis of such discontent, especially in developing countries experiencing rapid urbanization, like Chile.

This perspective, deeply embedded in contemporary discussions, especially due to significant contributions from the field of geography, has also been a recurring concern in political science. Classic works like that of Almond and Verba (1989) expressed a distinctive focus on territorial levels when exploring individuals' perceived competence in political participation. The authors emphasized that the perceived accessibility and responsiveness of local and national institutions varied and should be highlighted when examining civic culture.

> Adding these three facts together—local and national competence are related, local competence is more widespread than national, and local competence is related to the institutional availability of opportunities to participate on the local level—one has an argument in favor of the classic position that political participation on the local level plays a major role in the development of a competent citizenry. As many writers have argued, local government may act as a training ground for political competence. Where local government allows participation, it may foster a sense of competence that then spreads to the national level—a sense of competence that would have had a harder time developing had the individual's only involvement with government been with the more distant and inaccessible structures of the national government. To argue this point is to speculate beyond our data on national and local competence. But in a later chapter, we shall present data to the effect that the individual's belief in his ability to affect the government derives, at least in part, from opportunities to be influential within smaller authority structures such as the family, the school, and the place of work (Almond & Verba, 1989, p. 145).

2.4.3 Back to the Classics: The Politics of Resentment and Its Territorial Roots

Over a decade ago, research began to refocus on contextual elements that might help explain the rise of anti-establishment narratives. Katherine J. Cramer's work has become essential in understanding these dynamics (Cramer, 2012, 2016). Concentrating on the situation in Wisconsin (USA), Cramer aimed to study the impact of "group consciousness" on political preferences, which sometimes result in votes that can appear contrary to the voters' own interests. By examining the local context, her research suggests that in some areas, people possess a class and place-based identity intertwined with a perception of deprivation.

In essence, it is not just individual factors that lead people to adopt certain preferences but also an anti-elitist sentiment deeply rooted in their territory. This "rural consciousness" stems from the feeling that urban political elites show disdain and disrespect for rural residents and their unique lifestyles. This idea is central, as the experienced marginalization is attributed to a specific group (in this case, urban elites who hold power). Rural inhabitants often complain that decisions are made in urban areas by groups who do not share their interests and then merely communicate these decisions. Therefore, this consciousness leads to a propensity to limit the reach of a government perceived as "anti-rural", even if it adversely affects the population's particular interests.

So, what is rural consciousness? According to Cramer (2012), the concept encompasses several characteristics. For instance, it relates not only to the distribution of resources but also to the distribution of political power. It is anchored in a perception of distributive injustice against a hardworking group—marginalized rural sectors. This is also relevant because, as Cramer suggests, it results in a sense of ingroup pride that contrasts with a feeling of relative deprivation.

In these terms, and following Cramer's perspective, the concept is fundamentally linked to political attitudes usually associated with disaffection. According to Cramer, rural consciousness involves a feeling of low external efficacy, meaning the belief that the political system does not respond to the interests of rural residents. This is particularly interesting because the author encapsulates these feelings of neglect and abandonment through the classic concept of system responsiveness.

Following Cramer's insights, we observe that the propensity to embrace anti-elitist narratives is not solely attributable to individual factors but is also deeply influenced by the context in which people live. These conditions shape feelings, attitudes, and consequently, preferences.

However, recent researchers have suggested that the rise of anti-elitism and its connection to the territorial context extends beyond the urban-rural divide. For instance, in Norway, building on the previous work of Lipset and Rokkan (1967), researchers have examined the impact of the center-periphery cleavage on political trust, another variable commonly associated with disaffection. Utilizing a multilevel model with over 20,000 cases, they concluded that peripheral regional location, measured for example by distance from the political center, has a greater impact on political trust than the urban-rural divide (Stein et al., 2021).

Similarly, Ziblatt et al. (2023) also employ the center-periphery cleavage to explain the rise of the radical right in Germany. They argue that what is often perceived as the rural base of these radical right movements can actually be attributed to a different phenomenon: communities historically situated on the periphery in center-periphery conflicts that have shaped modern nation-states. Their innovative approach does not rely on a dichotomous distinction between center and periphery. Instead, they utilize classical literature on nonstandard linguistic dialects across different regions to identify peripheral areas through the prevalence of a "wealth of tongues", drawing on more than 725,000 geo-coded responses from a linguistic survey in Germany. Ultimately, the authors demonstrate that historically peripheral communities are more likely to vote for the radical right today.

The cases of Norway and Germany, among others, transcend the traditional urban-rural distinction while reflecting similar patterns. The feeling of abandonment is influenced not just by individual factors but by the specific contexts in which citizens live. In some places, even if they are not necessarily poor, residents feel more excluded, leading to stronger anti-elitist sentiments and "out-group resentment". These feelings make inhabitants more likely to support parties that challenge the elite, such as populist or radical right-wing movements. This argument underlies both center-periphery studies and those examining the rural-urban divide.

2.4.4 Geography of Discontent: Political, Historical, Geographical, and Economic Contexts

The phenomenon of contextual determinants of political resentment and discontent has also been explored from various perspectives and disciplines. In light of Donald Trump's rise and the growth of Euroscepticism—including Brexit—authors have suggested that processes of deindustrialization and recent economic decline have resulted in "places that don't matter", with inhabitants accumulating feelings of powerlessness, anger, and frustration, often manifested at the ballot box (Goodwin & Heath, 2016; Rodrigues & Meza, 2018).

The argument extends beyond individual factors like gender, age, socioeconomic level, and personal income in explaining the propensity to support anti-system narratives. To fully understand recent phenomena, authors suggest that it is crucial to consider the historical, geographical, institutional, economic, and political contexts of the places where citizens live. These elements collectively shape identities, preferences, and even the structure of feelings of individuals (McQuarrie, 2017).

In this context, several authors have introduced the concept of the "geography of discontent" (Los et al., 2017; Dijkstra et al., 2020), which has been pivotal in understanding the rise of nationalist, Eurosceptic, and far-right movements in Europe. For instance, in Italy, some authors have argued that the historical and socioeconomic differences between the south and the north were essential to explain the results in the 2018 Senate elections and the rise of different populist narratives, which are also related to the classical urban-rural divide and other local characteristics (Faggian et al., 2021). In the United Kingdom, various scholars argue that territorial conditions were crucial in explaining the propensity to choose the leave option in the 2016 Brexit referendum (Los et al., 2017; Alabrese et al., 2019). This needs to be understood within the context of territorial polarization reshaping British politics (Jennings & Stoker, 2019), influenced not just by economic considerations but also by cultural factors such as diversity, class structures, cosmopolitanism, and localism (Gordon, 2018; Jennings et al., 2016). In Austria, specifically in its capital city, Vienna, it has been shown that neighborhood characteristics were significant in explaining the triumph of right-wing populism. Authors argue that the socioeconomic conditions of neighborhoods are more important than individual socioeconomic conditions in explaining these phenomena (Essletzbichler & Forcher, 2022). This framework helps explain how specific places, influenced by their unique histories and economic trajectories, become fertile ground for political resentment and discontent. The "geography of discontent" emphasizes that it is not just personal circumstances but also the broader spatial and contextual factors that drive support for these movements.

Among many other things, this perspective invites us to reconsider the impact of economic realities on the emergence of anti-system projects and, specifically, nationalist and populist narratives. At the individual level, it is often assumed that these anti-establishment attitudes are not directly related to socioeconomic factors, but this changes when viewed through a spatial lens, especially considering the trajectory of specific places. In this context, the thesis of Rodriguez-Pose et al. (2024) becomes particularly relevant. According to the authors, there are places that progress while coexisting with others that, even if they have reached a certain level of development, are mired in what they term a "regional development trap". Specifically, they describe regions experiencing stagnation or outright regression in various related variables such as employment and growth. This aligns with our previous discussions: prolonged periods in these traps lead inhabitants to feel dissatisfaction and unrest, which then becomes evident in their electoral preferences.

Although it is already well-established that economic decline, deindustrialization, and demographic decline (among other factors) can lead to anti-system narratives (McCann, 2020; Rodríguez-Pose et al., 2021; Dijkstra et al., 2020), the authors offer a more holistic perspective that aims to integrate various factors. Their findings are particularly compelling because they transcend mere vulnerability or deprivation. Even those living in relatively developed areas (middle or high-income regions) show a greater propensity to vote for Eurosceptic narratives when facing periods of stagnation or decline. This propensity becomes stronger the longer the duration of the trap.

Up to this point, we can understand this as a complex process interconnected with various phenomena. It is not merely about vulnerability or urban-rural divides; its determinants seem to interact with diverse elements. For instance, some authors have suggested the potential impact of certain European policies related to the green transition. According to Rodriguez-Pose et al. (2023), the promoted measures would affect different regions of Europe in varying ways, with less developed and rural territories in southern and eastern Europe being more exposed to changes than others.

While much of the literature focuses on Europe, significant contributions have extended these findings to other developed countries, such as the United States. McQuerry (2017), for example, suggests that Donald Trump's rise can be explained by a disconnect between national parties and regional experiences of decline. Rodriguez-Pose et al. (2021), on the other hand, argue that areas experiencing economic and demographic decline are more prone to following populist narratives.

To assess these populist narratives, they consider not only the 2016 Trump vote but also the margin between Trump's support and that of Mitt Romney four years earlier. Using county-level data, they conclude that the effect of economic and demographic decline is even more pronounced in areas with higher social capital.

2.4.5 Concluding Observations

As we can observe, these are just some of the significant advancements made in recent years regarding the territorial and contextual elements that could explain discontent and the feeling of abandonment—or the politics of resentment. However, in empirical terms, studies have often focused primarily on the behavioral component of political engagement. The literature tends to identify appropriate socioeconomic or demographic factors to explain citizen behavior at the polls. Along the way, it often assumes the existence of certain attitudes, usually associated with the frustration and anger that arise from being "left behind". This concern persists in the literature, especially in studies related to the geography of discontent. For Becker et al. (2017), the success of Brexit was precisely due to these sensations and attitudes. According to the authors, that vote represented an opportunity for the excluded to "express their anger" toward an inefficient ruling class. Similarly, other scholars acknowledge the need to study these underlying elements that could explain voting behavior, which, as noted, is often linked to feelings of powerlessness and abandonment discussed by Cramer (2012; 2016), Fitzgerald (2011), and others.

The challenge, then, is to move from the "geography of voting" to a more specific "geography of discontent", understanding the contextual elements that could explain these harmful feelings for democratic coexistence. By doing so, we are not only extending the literature on the geography of discontent to more attitudinal aspects but also opening new avenues regarding the potential role digital platforms could play in integrating these left-behind territories.

References

Abramson, P. (1983). *Political attitudes in America: Formation and change*. W.H: Freeman & Co Ltd.

Acock, A. C., & Clarke, H. D. (1990). Alternative measures of political efficacy: Models and means. *Quality and Quantity, 24*(1), 87–105. https://doi.org/10.1007/BF00221386

Alabrese, E., Becker, S. O., Fetzer, T., & Novy, D. (2019). Who voted for Brexit? Individual and regional data combined. *European Journal of Political Economy, 56*, 132–150. https://doi.org/10.1016/j.ejpoleco.2018.08.002

Almond, G. A., & Verba, S. (1989). *The civic culture. Political attitudes and democracy in five nations*. Sage.

Andrews, R., Entwistle, T., & Guarneros-Meza, V. (2019). Local government size and political efficacy: Do citizen panels make a difference? *International Journal of Public Administration, 42*(8), 664–676. https://doi.org/10.1080/01900692.2018.1499774

Balch, G. I. (1974). Multiple indicators in survey research: The concept "Sense of Political Efficacy". *Political Methodology,1*(2), 1–43. http://www.jstor.org/stable/25791375

Becker, S. O., Fetzer, T., & Novy, D. (2017). Who voted for Brexit? A comprehensive district-level analysis. *Economic Policy, 32*(92), 601–650. https://doi.org/10.1093/epolic/eix012

Best, S. J., & Krueger, B. S. (2005). Analyzing the representativeness of Internet political participation. *Political Behavior, 27*(2), 183–216. https://doi.org/10.1007/s11109-005-3242-y

Bonfadelli, H. (2002). The internet and knowledge gap: A theoretical and empirical investigation. *European Journal of Communication, 17*(1), 65–84.

Booth, J. A., & Seligson, M. A. (2009). *The legitimacy puzzle in Latin America: Political support and democracy in eight nations*. Cambridge University Press.

Boulianne, S., Koc-Michalska, K., & Bimber, B. (2020). Right-wing populism, social media and echo chambers in Western democracies. *New Media & Society, 22*(4), 683–699. https://doi.org/10.1177/1461444819893983

Boulianne, S., Oser, J., & Hoffmann, C. P. (2023). Powerless in the digital age? A systematic review and meta-analysis of political efficacy and digital media use. *New Media & Society, 14614448231176520*. https://doi.org/10.1177/14614448231176519

Campbell, A., Gurin, G., & Miller, W. E. (1954). *The voter decides*. Row, Peterson & Co.

Cardoso, F., & Faletto, E. (1976). *Dependencia y desarrollo en América Latina*. Siglo XXI Editores.

Carrara, S. (2012). Towards e-ECIs? European participation by online pan-European mobilization. *Perspectives on European Politics and Society, 13*(3), 352–369. https://doi.org/10.1080/15705854.2012.702578

Chan, M., Chen, H.-T., & Lee, F. L. F. (2016). Examining the roles of mobile and social media in political participation: A cross-national analysis of three Asian societies using a communication mediation approach. *New Media & Society, 19*(12), 2003–2021. https://doi.org/10.1177/1461444816653190

Chan, M., Chen, H.-T., & Lee, F. L. F. (2018). Examining the roles of political social network and internal efficacy on social media news engagement: A comparative study of six Asian countries. *The International Journal of Press/Politics, 24*(2), 127–145. https://doi.org/10.1177/1940161218814480

Chen, C., Bai, Y., & Wang, R. (2019). Online political efficacy and political participation: A mediation analysis based on the evidence from Taiwan. *New Media & Society, 21*(8), 1667–1696. https://doi.org/10.1177/1461444819828718

Coverse, P. E. (1972). Change in the American electorate. In A. Campbell & P. E. Converse (Eds.), *The human meaning of social change*. Russell Sage Foundation.

Craig, S. C., & Maggiotto, M. A. (1982). Measuring political efficacy. *Political Methodology,8*(3), 85–109. Retrieved April 27, 2023, from http://www.jstor.org/stable/25791157

Craig, S. C., Niemi, R. G., & Silver, G. E. (1990). Political efficacy and trust: A report on the NES pilot study items. *Political Behavior,12*(3), 289–314. http://www.jstor.org/stable/586303

Craig, S. C. (1980). The mobilization of political discontent. *Political Behavior, 2*(2), 189–209. https://doi.org/10.1007/BF00989890

Cramer, K. J. (2016). *The politics of resentment. Rural consciousness in wisconsin and the rise of Scott Walker*. The Chicago University Press.

Cramer, K. J. (2012). Putting inequality in its place: Rural consciousness and the power of perspective (2012/07/30). *American Political Science Review, 106*(3), 517–532. https://doi.org/10.1017/S0003055412000305

Dahl, R. A. (1967). The city in the future of democracy. *The American Political Science Review, 61*(4), 953–970. https://doi.org/10.2307/1953398

Dahl, R. A., & Tufte, E. R. (1973). *Size and democracy*. Stanford University Press.

Di Maggio, P., Hargittai, E., Celeste, C., & Shafer, S. (2004). Digital inequality. In K. M. Neckerman (Ed.), *Social inequality*. The Russell Sage Foundation.

Di Palma, G. (1970). *Apathy and participation. Mass politics in western societies*. The Free Press.

Di Palma, G. (1969). Disaffection and participation in western democracies: The role of political oppositions. *The Journal of Politics, 31*(4), 984–1010. https://doi.org/10.2307/2128355

References

Diehl, T., Weeks, B. E., & Gil de Zúñiga, H. (2016). Political persuasion on social media: Tracing direct and indirect effects of news use and social interaction. *New Media & Society, 18*(9), 1875–1895. https://doi.org/10.1177/1461444815616224

Dijkstra, L., Poelman, H., & Rodríguez-Pose, A. (2020). The geography of EU discontent. *Regional Studies, 54*(6), 737–753. https://doi.org/10.1080/00343404.2019.1654603

Easton, D. (1975). A re-assessment of the concept of political support. *British Journal of Political Science,5*(4), 435–457. http://www.jstor.org/stable/193437

Easton, D. (1976). Theoretical approaches to political support. *Canadian Journal of Political Science/Revue canadienne de science politique,9*(3), 431–448. http://www.jstor.org/stable/3230608

Easton, D., & Dennis, J. (1967). The child's acquisition of regime norms: Political efficacy. *The American Political Science Review, 61*(1), 25–38. https://doi.org/10.2307/1953873

Espinal, R., & Zhao, S. (2015). Gender gaps in civic and political participation in Latin America. *Latin American Politics and Society, 57*(1), 123–138. https://doi.org/10.1111/j.1548-2456.2015.00262.x

Essletzbichler, J., & Forcher, J. (2022). "Red Vienna" and the rise of the populist right. *European Urban and Regional Studies, 29*(1), 126–141. https://doi.org/10.1177/09697764211031622

Faggian, A., Modica, M., Modrego, F., & Urso, G. (2021). One country, two populist parties: Voting patterns of the 2018 Italian elections and their determinants. *Regional Science Policy & Practice, 13*(2), 397–413. https://doi.org/10.1111/rsp3.12391

Fierro, P., Aroca, P., & Navia, P. (2020). How people access the Internet and the democratic divide: Evidence from the Chilean region of Valparaiso 2017, 2018 and 2019. *Technology in Society, 63*, 101432. https://doi.org/10.1016/j.techsoc.2020.101432

Fierro, P., Aroca, P., & Navia, P. (2023). Political disaffection in the digital age: The use of social media and the gap in internal and external efficacy. *Social Science Computer Review, 41*(5), 1857–1876. https://doi.org/10.1177/08944393221087940

Fierro, P., Aroca, P., & Navia, P. (2023). The center-periphery cleavage and online political efficacy (OPE): Territorial and democratic divide in Chile, 2018–2020. *New Media & Society, 25*(6), 1335–1353. https://doi.org/10.1177/14614448211019303

Finkel, S. E. (1985). Reciprocal effects of participation and political efficacy: A panel analysis. *American Journal of Political Science, 29*(4), 891–913. https://doi.org/10.2307/2111186

Finkel, S. E. (1987). The effects of participation on political efficacy and political support: Evidence from a West German Panel. *The Journal of Politics, 49*(2), 441–464. https://doi.org/10.2307/2131308

Fischer, C. S. (1975). The city and political psychology (2014/08/01). *American Political Science Review, 69*(2), 559–571. https://doi.org/10.2307/1959086

Fitzgerald, J., & Lawrence, D. (2011). Local cohesion and radical right support: The case of the Swiss People's Party. *Electoral Studies, 30*(4), 834–847. https://doi.org/10.1016/j.electstud.2011.08.004

Friemel, T. N. (2014). The digital divide has grown old: Determinants of a digital divide among seniors. *New Media & Society, 18*(2), 313–331. https://doi.org/10.1177/1461444814538648

Gamson, W. A. (1968). *Power and discontent*. Dorsey Press.

Geurkink, B., Zaslove, A., Sluiter, R., & Jacobs, K. (2019). Populist attitudes, political trust, and external political efficacy: Old wine in new bottles? *Political Studies, 68*(1), 247–267. https://doi.org/10.1177/0032321719842768

Gibson, R. K. (2009). New media and the revitalisation of politics. *Representation, 45*(3), 289–299. https://doi.org/10.1080/00344890903129566

Goodwin, M. J., & Heath, O. (2016). *Brexit vote explained: Poverty, low skills and lack of opportunities*. Joseph Rowntree Foundation. https://www.jrf.org.uk/report/brexit-vote-explained-poverty-low-skills-and-lack-opportunities?gclid=EAIaIQobChMIqeyL7ICY1gIVhhobCh0IrwaBEAAYASAAEgIhdvD_BwE

Gordon, I. R. (2018). In what sense left behind by globalisation? Looking for a less reductionist geography of the populist surge in Europe. *Cambridge Journal of Regions, Economy and Society, 11*(1), 95–113. https://doi.org/10.1093/cjres/rsx028

Graham, M., De Sabbata, S., Straumann, R. K., & Ojanperä, S. (2018). Uneven digital geographies... and Why they matter. In K. Orangotango+ (Ed.), *This is not an Atlas* (transcript).

Hansen, K. M., & Tue Pedersen, R. (2014). Campaigns matter: How voters became knowledgeable and efficacious during election campaigns. *Political Communication, 31*(2), 303–324. https://doi.org/10.1080/10584609.2013.815296

Hargittai, E., & Hinnant, A. (2008). Digital inequality. *Communication Research, 35*(5), 602–621. https://doi.org/10.1177/0093650208321782

Hirzalla, F., van Zoonen, L., & de Ridder, J. (2011). Internet use and political participation: reflections on the mobilization/normalization controversy. *The Information Society, 27*(1), 1–15. https://doi.org/10.1080/01972243.2011.534360

Hoffmann, C. P., & Lutz, C. (2019). Digital divides in political participation: The mediating role of social media self-efficacy and privacy concerns. *Policy & Internet, n/a*(n/a). https://doi.org/10.1002/poi3.225

Jennings, W., & Stoker, G. (2019). The divergent dynamics of cities and towns: Geographical polarisation and Brexit. *The Political Quarterly, 90*(S2), 155–166. https://doi.org/10.1111/1467-923X.12612

Jennings, W., Stoker, G., & Twyman, J. (2016). The dimensions and impact of political discontent in Britain. *Parliamentary Affairs, 69*(4), 876–900. https://doi.org/10.1093/pa/gsv067

Kenski, K., & Stroud, N. J. (2006). Connections between Internet use and political efficacy, knowledge, and participation. *Journal of Broadcasting & Electronic Media, 50*(2), 173–192. https://doi.org/10.1207/s15506878jobem5002_1

Kneuer, M., & Datts, M. (2020). E-democracy and the matter of scale. Revisiting the democratic promises of the Internet in terms of the spatial dimension. *Politische Vierteljahresschrift,61*(2), 285–308. https://doi.org/10.1007/s11615-020-00250-6

Lane, R. (1959). *Political life: Why people get involved in politics.*

Lipset, S. M., & Rokkan, S. (1967). Cleavage structures, party systems, and voter alignments: an introduction. In S. M. Lipset & S. Rokkan (Eds.), *Party systems and voter alignments; cross-national perspectives* (pp. 1–64). The Free Press.

Los, B., McCann, P., Springford, J., & Thissen, M. (2017). The mismatch between local voting and the local economic consequences of Brexit. *Regional Studies, 51*(5), 786–799. https://doi.org/10.1080/00343404.2017.1287350

Luna, J. P. (2016). Chile's crisis of representation. *Journal of Democracy, 27*(3), 129–138. https://doi.org/10.1353/jod.2016.0046

Maldonado Hernández, G. (2013). Desapego político y desafección institucional en México ¿Desafíos para la calidad de la democracia? *Política y gobierno; Tematico.* http://www.politicaygobierno.cide.edu/index.php/pyg/article/view/1053

Martin, J. D., Martins, R. J., & Naqvi, S. (2018). Media use predictors of online political efficacy among Internet users in five Arab countries. *Information, Communication & Society, 21*(1), 129–146. https://doi.org/10.1080/1369118X.2016.1266375

McCann, P. (2020). Perceptions of regional inequality and the geography of discontent: Insights from the UK. *Regional Studies, 54*(2), 256–267. https://doi.org/10.1080/00343404.2019.1619928

McDonnell, J. (2020). Municipality size, political efficacy and political participation: A systematic review. *Local Government Studies, 46*(3), 331–350. https://doi.org/10.1080/03003930.2019.1600510

McKay, L. (2019). 'Left behind' people, or places? The role of local economies in perceived community representation. *Electoral Studies, 60,* 102046. https://doi.org/10.1016/j.electstud.2019.04.010

McQuarrie, M. (2017). The revolt of the Rust Belt: Place and politics in the age of anger. *The British Journal of Sociology, 68*(S1), S120–S152. https://doi.org/10.1111/1468-4446.12328

Miller, W. E., Miller, A. H., & Schneider, E. J. (1980). *American National Election Studies Data Sourcebook, 1952–1978.* Harvard University Press.

Min, S.-J. (2010). From the digital divide to the democratic divide: Internet skills, political interest, and the second-level digital divide in political Internet use. *Journal of Information Technology & Politics, 7*(1), 22–35. https://doi.org/10.1080/19331680903109402

Montero, J. R., Sanz, A., & Navarrete, R. M. (2016). La democracia en tiempos de crisis: Legitimidad, descontento y desafección en España. In J. L. Cascajo Castro & A. M. de la Vega (Eds.), *Participaci ó n, representaci ó n y democracia* (Primera). Tirant Lo Blanch.

Montero, J. R., Gunther, R., & Torcal, M. (1997). Democracy in Spain: Legitimacy, discontent, and disaffection. *Studies in Comparative International Development, 32*(3), 124–160. https://doi.org/10.1007/BF02687334

Morrell, M. E. (2003). Survey and experimental evidence for a reliable and valid measure of internal political efficacy. *Public Opinion Quarterly, 67*(4), 589–602. https://doi.org/10.1086/378965

Morris, D. S., & Morris, J. S. (2013). Digital inequality and participation in the political process. *Social Science Computer Review, 31*(5), 589–600. https://doi.org/10.1177/0894439313489259

Navarrete, B., & Tricot, V. (2021). Introduction: Social outbreak and political representation in Latin America. In B. Navarrete & V. Tricot (Eds.), *The social outburst and political representation in Chile* (pp. 1–10). Springer International Publishing. https://doi.org/10.1007/978-3-030-70320-2_1

Newton, K. (1982). Is small really so beautiful? Is Big really so ugly? Size, effectiveness, and democracy in local government. *Political Studies, 30*(2), 190–206. https://doi.org/10.1111/j.1467-9248.1982.tb00532.x

Nie, N. H., Powell, G. B., & Prewitt, K. (1969). Social structure and political participation: Developmental relationships, Part I. *The American Political Science Review, 63*(2), 361–378. https://doi.org/10.2307/1954694

Niemi, R. G., Craig, S. C., & Mattei, F. (1991). Measuring internal political efficacy in the 1988 national election study. *The American Political Science Review, 85*(4), 1407–1413. https://doi.org/10.2307/1963953

Norris, P. (2001). *Digital divide?: Civic engagement*. Information poverty and the Internet Worldwide (First): Cambridge University Press.

Nye, J., Zelikow, P. D., & King, D. C. (1997). *Why people don't trust government*. Harvard University Press.

OECD. (2020). *Digital government in Chile—Improving public service design and delivery*.

Offe, C. (2006). Political disaffection as an outcome of institutional practices? Some post-Tocquevillean speculations. In M. Torcal & J. R. Montero (Eds.), *Political disaffection in contemporary democracies*. Routledge.

Oliver, J. E. (2000). City size and civic involvement in metropolitan America. *The American Political Science Review, 94*(2), 361–373. https://doi.org/10.2307/2586017

Oser, J., Grinson, A., Boulianne, S., & Halperin, E. (2022). How political efficacy relates to online and offline political participation: A multilevel meta-analysis. *Political Communication, 39*(5), 607–633. https://doi.org/10.1080/10584609.2022.2086329

Putnam, R. D. (1995). Tuning in, tuning out: The strange disappearance of social Capital in America. *PS: Political Science and Politics,28*(4), 664–683. https://doi.org/10.2307/420517

Putnam, R. D. (2000). *Bowling alone: The collapse and revival of American community*. Simon & Schuster.

Rodrigues, M., & Meza, O. D. (2018). "Is there anybody out there?" Political implications of a territorial integration. *Journal of Urban Affairs, 40*(3), 426–440. https://doi.org/10.1080/07352166.2017.1360732

Rodríguez-Pose, A. (2020). The rise of populism and the revenge of the places that don't matter. *LSE Public Policy Review,1*(1). https://doi.org/10.31389/lseppr.4

Rodríguez-Pose, A., & Bartalucci, F. (2023). The green transition and its potential territorial discontents. *Cambridge Journal of Regions, Economy and Society*, rsad039. https://doi.org/10.1093/cjres/rsad039

Rodríguez-Pose, A., Lee, N., & Lipp, C. (2021). Golfing with Trump. Social capital, decline, inequality, and the rise of populism in the US. *Cambridge Journal of Regions, Economy and Society,14*(3), 457–481. https://doi.org/10.1093/cjres/rsab026

Rodríguez-Pose, A. (2018). The revenge of the places that don't matter (and what to do about it). *Cambridge Journal of Regions, Economy and Society, 11*(1), 189–209. https://doi.org/10.1093/cjres/rsx024

Rodríguez-Pose, A., Dijkstra, L., & Poelman, H. (2024). The geography of EU discontent and the regional development trap. *Economic Geography, 100*(3), 213–245. https://doi.org/10.1080/00130095.2024.2337657

Saglie, J., & Vabo, S. I. (2009). Size and e-democracy: Online participation in Norwegian local politics. *Scandinavian Political Studies, 32*(4), 382–401. https://doi.org/10.1111/j.1467-9477.2009.00235.x

Sasaki, F. (2016). Online Political Efficacy (OPE) as a reliable survey measure of political empowerment when using the Internet. *Policy & Internet, 8*(2), 197–214. https://doi.org/10.1002/poi3.114

Sasaki, F. (2017). Does Internet use provide a deeper sense of political empowerment to the less educated? *Information, Communication & Society, 20*(10), 1445–1463. https://doi.org/10.1080/1369118X.2016.1229005

Scheufele, D. A., & Nisbet, M. C. (2002). Being a citizen online. *Harvard International Journal of Press/Politics, 7*(3), 55–75. https://doi.org/10.1177/1081180X0200700304

Schlozman, K. L., Verba, S., & Brady, H. E. (2010). Weapon of the strong? Participatory inequality and the Internet. *Perspectives on Politics,8*(2), 487–509. http://www.jstor.org/stable/25698614

Shingles, R. D. (1981). Black consciousness and political participation: The missing link. *The American Political Science Review, 75*(1), 76–91. https://doi.org/10.2307/1962160

Stein, J., Buck, M., & Bjørnå H. (2021). The centre-periphery dimension and trust in politicians: The case of Norway. *Territory, Politics, Governance, 9*(1), 37–55. https://doi.org/10.1080/21622671.2019.1624191

Stoycheff, E., Nisbet, E. C., & Epstein, D. (2016). Differential effects of capital-enhancing and recreational internet use on citizens' demand for democracy. *Communication Research,0093650216644645*. https://doi.org/10.1177/0093650216644645

Torcal, M., & Lago, I. (2006). Political participation, information, and accountability: Some consequences of political disaffection in new democracies. In M. Torcal & J. R. Montero (Eds.), *Political disaffection in contemporary democracies. Social capital, institutions, and politics* (pp. 308–332). Rutledge.

Torcal, M., & Montero, J. R. (2006a). Political disaffection in comparative perspective. In M. Torcal & J. R. Montero (Eds.), *Political disaffection in contemporary democracies. Social capital, institutions, and politics* (pp. 3–20). Rutledge.

Torcal, M. (2006). Desafección institucional e historia democrática en las nuevas democracias. *Revista SAAP, 2*(3), 591–634.

Torcal, M., & Montero, J. R. (2006). *Political disaffection in contemporary democracies*. Routledge.

Valentino, N. A., Gregorowicz, K., & Groenendyk, E. W. (2009). Efficacy, emotions and the habit of participation. *Political Behavior,31*(3), 307–330. http://www.jstor.org/stable/40587287

van Deursen, A. J., & van Dijk, J. A. (2014). The digital divide shifts to differences in usage. *New Media & Society, 16*(3), 507–526. https://doi.org/10.1177/1461444813487959

Wagner, K. M., Gray, T. J., & Gainous, J. (2017). Digital information consumption and external political efficacy in Latin America: Does institutional context matter? *Journal of Information Technology & Politics, 14*(3), 277–291. https://doi.org/10.1080/19331681.2017.1337601

Wallerstein, I. (1974). *The modern World System I: Capitalist agriculture and the origins of the European wolrd-economy in the sixteenth century*. Academic Press.

Wilson, K. R., Wallin, J. S., & Reiser, C. (2003). Social stratification and the digital divide. *Social Science Computer Review, 21*(2), 133–143. https://doi.org/10.1177/0894439303021002001

Wirth, L. (1938). Urbanism as a way of life. *American Journal of Sociology,44*(1), 1–24. http://www.jstor.org/stable/2768119

References

Zhang, W., Johnson, T. J., Seltzer, T., & Bichard, S. L. (2010). The revolution will be networked: The influence of social networking sites on political attitudes and behavior. *Social Science Computer Review, 28*(1), 75–92. https://doi.org/10.1177/0894439309335162

Ziblatt, D., Hilbig, H., & Bischof, D. (2023). Wealth of tongues: Why peripheral regions vote for the radical right in Germany (2023/10/20). *American Political Science Review, 1–17*. https://doi.org/10.1017/S0003055423000862

Chapter 3
Methodological Approach

Abstract This chapter introduces a method encompassing three political attitudes commonly linked to disaffection: internal political efficacy, external political efficacy, and political interest. We investigate their interactions with social media use for civic engagement and the type of device used to access the Internet. Additionally, we incorporate a contemporary measure, online political efficacy, designed specifically for the digital sphere to evaluate whether digital platforms can empower individuals typically marginalized in political discussions. To achieve these goals, we analyzed survey data from the Chilean non-profit P!ensa Foundation, covering 11,574 face-to-face cases collected between 2017 and 2023. Using Cronbach's Alpha, we validated factors related to political attitudes and social media use, distinguishing between civic use (political and informational) and other use (social interaction and entertainment). Structural equation modeling (SEM) was employed to explore these factors and their interactions. While most studies emphasize individual or institutional aspects, recent "Geography of Discontent" literature highlights territorial disparities. This study reexamines civic engagement by integrating territorial trajectories and experiences across regions, cities, or neighborhoods.

Keywords Survey design · Structural equation model · Spatial autocorrelation · Global south · Political attitudes

3.1 Questionnaire Design

3.1.1 Assessing Political Attitudes Usually Associated with Political Disaffection

As we have mentioned, our work aims to contribute to the literature on the contextual elements that explain discontent, focusing on an attitudinal rather than behavioral approach (such as voting patterns). We seek to shed light on what is often an assumption—those feelings of abandonment, frustration, and powerlessness.

Political attitudes often serve as explanatory variables when examining behaviors, extending to digital engagement as well. The political or informational use of social media platforms, for instance, is usually explained by different attitudes and sentiments toward politics and political issues. In our studies, we have prioritized three specific attitudes commonly associated with political disaffection and, in our terms, those feelings of exclusion we aim to encompass: internal political efficacy, external political efficacy, and political interest.

As previously stated, the notion of political efficacy traces back to Campbel et al. (1954), who characterized it as the sentiment that an individual's political actions can significantly influence the political process, underscoring the value of fulfilling one's civic responsibilities. Initially conceived as a singular concept, subsequent scholarly discourse suggested a more nuanced understanding. Coverse (1972) introduced a distinction between an individual's "sense of political competence" and their "faith in system responsiveness". Echoing this, Balch (1974) crystallized the now-classic bifurcation into internal and external efficacy. Internal efficacy reflects an individual's grasp of public affairs and their confidence in their ability to meaningfully engage in political processes. Conversely, external efficacy encompasses perceptions of how responsive political systems and institutions are to citizen advocacy.

As mentioned in previous chapters, both internal and external efficacy have been used to measure different aspects of political disaffection. Following the distinction made by Torcal and Montero (2006) for the specific case of Latin America, internal efficacy—i.e., the sense of self-competence—could be associated with political disengagement, while external efficacy—i.e., system responsiveness beliefs—could be associated with institutional disaffection.

Notwithstanding the general consensus in academic circles regarding this bifurcation, certain studies exploring the nexus between attitudes and social media usage have either treated political efficacy as a singular entity or spotlighted only one of its facets. Additionally, there's an ongoing debate regarding the measurement of these variables. While some scholars advocate for multi-dimensional scales, numerous studies in communications opt for a singular measure for each facet.

Navigating these complexities, our proposal incorporates five questions to gauge internal efficacy and four for external efficacy. Crucially, our choice of phrasing draws inspiration from Hansen and Tue Pedersen (2014), ensuring coherence and robustness in our approach (see Tables 3.1 and 3.2). Hansen and Tue Pedersen specifically worked with these sets of items to understand if political campaigns may affect these specific attitudes. Their results are relevant and pertinent to our purpose because they showed that political communication indeed positively influences political attitudes that seem to be more permanent and less volatile to specific events.

When it comes to political interest, a segment of scholars proposes its measurement in alignment with exposure for political ends. However, this method leans heavily on behavioral indicators rather than attitudinal ones, a distinction underscored by Stromback and Shehata (2010). In light of this, and in line with a significant body of academic recommendations, our study measures political interest through self-reported interest, as detailed in Table 3.3.

3.1 Questionnaire Design

Table 3.1 Questions on internal political efficacy

Question: How much do you agree with the following statements?	Variable
Sometimes politics is so complicated that a person like me cannot really understand what is going on	inef1
Generally speaking, I do not find it that difficult to take a stand on political issues	inef2
When politicians debate economic policy, I only understand a small part of what they are talking about	inef3
Citizens like me are qualified to participate in political discussions	inef4
Citizens like me have opinions on politics that are worth listening to	inef5

Note The answers were measured on a 5-point Likert scale

Table 3.2 Questions on external political efficacy

Question: How much do you agree with the following statements?	Variable
Politicians do not really care what the voters think	exef1
Usually you can trust the political leaders to do what is best for the country	exef2
The politicians waste a lot of the taxpayer's money	exef3
Citizens like me do not have any influence on the decisions of Parliament and Government	exef4

Note The answers were measured on a 5-point Likert scale

Table 3.3 Questions on political interest

Question: How interested are you in...?	Variable
Politics	polint1
Situations that occur in your city	polint2
Situations that occur in your region	polint3
Situations that occur in the country	polint4
Situations that occur in the world	polint5

Note The answers were measured on a 4-point Likert scale

Various researchers have adopted different approaches to studying political attitudes. In the case of political interest, some argue that using self-reported answers may introduce bias, especially in countries where disconnection from public affairs is socially frowned upon, as is often the case in Chile. In such contexts, there might be incentives to exaggerate one's interest in public affairs. Consequently, some scholars suggest that a plausible alternative is to use questions focused on media exposure or political content consumption (Lupia & Philpot, 2005; Oskarson, 2007), aiming to measure the respondent's real interest more "objectively". The downside of this approach, however, is that it shifts focus away from attitudinal aspects and more toward specific actions of individuals. Therefore, we have opted to continue using self-reported values.

Another interesting element is our choice to use a set of questions that provide more information than a single-item measure for all attitudes. This will be elaborated on in the following sections, but it is a crucial decision as it assumes the existence of latent variables—i.e., the attitudes we aim to capture—that are ultimately measured by specific observed variables—i.e., the set of questions used. We believe this approach is consistent with recommendations commonly made in the literature on political psychology.

3.1.2 Internet and Digital Platforms in Integrating Marginalized Areas

In addition to identifying the contextual elements that explain political discontent, our research also focuses on the role of the Internet and digital platforms in integrating areas that are often marginalized from public affairs. This approach is interesting as it compels us to extend our understanding of the digital divide—and specifically the democratic divide—by incorporating this territorial dimension.

To achieve this, we concentrate on two different aspects: firstly, the various ways social platforms—specifically social media—are used; and secondly, a special measure of political efficacy specifically designed for the online world.

3.1.2.1 Measuring Social Media Usage: From Entertainment to Engagement

One of the challenges researchers face is the lack of uniformity in defining and measuring social media usage. Consequently, some scholars argue that the varied findings surrounding social media's democratic implications stem partly from these divergent conceptual frameworks and metrics employed (Chan et al., 2016; Hyun & Kim, 2015).

Taking cues from influential studies on the nexus between social media and civic engagement (Diehl et al., 2016; Gil de Zúñiga et al., 2014; Yang & DeHart, 2016), our research harnesses seven distinct questions to assess four specific types of social media use: social interaction, entertainment, political engagement, and information-seeking. The specifics of each of these questions are articulated in Table 3.4.

3.1.2.2 The New Measure of Online Political Efficacy

Unlike the classic concept, Online Political Efficacy does not necessarily distinguish between the internal component (self-competence) and the external component (system-responsiveness beliefs). Instead, it functions as a general concept that measures the sense that, through the Internet, one's political actions are worthwhile and create change.

3.1 Questionnaire Design

Table 3.4 Questions on different types of social media use

Question: How interested are you in...?	Variable	
Social interaction	Stay in contact with friends and family	use1
Social interaction	Meet people who share my interests	use2
Entertainment	View videos and photos for entertainment	use3
Political expression	Share photos or videos about political issues	use4
Political expression	Share my opinion about political ideas	use5
Political expression	Publicize my experience in political or electoral activities	use6
Information and news	Get information about current events and public affairs	use7

Note The answers were measured on a 7-point Likert scale

Table 3.5 Questions on online political efficacy

Question: How interested are you in...?	Variable
By using the Internet, people like you can have more political power	ope1
By using the Internet, people like you can have more say in what the government does	ope2
By using the Internet, people like you can better understand politics	ope3
By using the Internet, public officials will care more about what people like me think	ope4

Note The answers were measured on a 5-point Likert scale

This measure has been widely used over the past five years in various contexts with very good results. For this project, we used the same set of questions proposed by the World Internet Project, consisting of four questions measured on a 5-point Likert scale. Those specific questions are shown in Table 3.5.

3.1.3 Additional Remarks

This chapter focuses on the role of territorial context in shaping the underlying feelings and elements that influence political behaviors such as voting or abstention. As such, political attitudes become an essential component of our analysis.

The decisions we have made in our approach are based on the application of concepts that have been traditionally used in studying civic engagement, particularly in the context of political disaffection. These political attitudes are instrumental in examining topics frequently discussed in the literature on space and politics, such as the feeling of abandonment experienced by residents of "places left behind", which we have suggested is associated with a decline in system responsiveness beliefs—i.e., external efficacy.

Furthermore, the political attitudes included in our questionnaires will enable us to propose new and innovative approaches, building on seminal works. For instance, political efficacies, collectively and particularly the gaps between self-competence beliefs and system responsiveness beliefs, can be highly useful for studying frustration and resentment in politics—phenomena often linked to spatial inequities and the feeling of being left behind.

Ultimately, the goal of this study is to shift from an analysis of the geography of voting to the geography of discontent, necessitating the inclusion of subjective and attitudinal elements.

3.2 The Analysis

3.2.1 Data Sources and Collection

In our pursuit of robust, actionable insights, we utilized a survey dataset provided by a well-regarded Chilean non-profit organization, the P!ensa Foundation. This survey, carried out by the global research firm Ipsos and the Chilean polling company Datavoz-Statcom, offers a wealth of rich, in-depth responses. The data is publicly available through the Harvard Dataverse repository. Permission to conduct interviews for academic research purposes was obtained from all respondents, who were fully informed about the objectives of this investigation and how their responses would be used.

A closer look at the survey specifics:

- **Scope and Sample Size**: Across seven consecutive years—2017, 2018, 2019, 2020, 2021, 2022, and 2023—1650 face-to-face interviews were conducted annually, summing up to a cumulative total of 11,574 interviews.
- **Demographics:** The respondents were all adults, aged 18 and above.
- **Geographical Coverage**: The survey's geographical purview was strategic, focusing on the 10 most populous municipalities in the Valparaíso region of Chile. Based on 2017 census data, these municipalities collectively account for a sizable 72% of the region's entire population.
- **Survey Timeframes**:
  - 2017: Conducted from March 18 to April 11, this round of interviews took place eight months before Chile's significant national elections in November, where both the presidency and legislative bodies were contested.
  - 2018: Running from May 11 through June 11, this wave of data collection followed soon after the inauguration of the newly elected government, just two months into their term.
  - 2019: This survey was administered between April 15 and May 22.
  - 2020: The fieldwork occurred between February 17 and April 20. Due to mobility restrictions and social distancing policies resulting from the COVID-19

3.2 The Analysis

Table 3.6 Descriptive statistics of the sample

Variable	Observations	Mean	Std. dev.	Min	Max
Gender (female = 1)	11574	0.6	0.49	0	1
Age	11574	47.6	17.9	18	96
SES	11467	2.9	1.07	1	5
Right	11574	0.12	0.33	0	1
Center	11574	0.1	0.3	0	1
Left	11574	0.18	0.39	0	1
Access to Internet from home	11556	0.76	0.42	0	1
Access to Internet from work	10559	0.42	0.49	0	1
Access to Internet from via mobile	11548	0.81	0.39	0	1

pandemic, 1413 in-person interviews were supplemented with 237 telephone interviews to complete the sample.
– 2021: Conducted between July 7 and September 13, this round of interviews concluded over two months before the national elections in November, where both the presidency and legislative bodies were up for a vote.
– 2022: This survey cycle took place from May 31 to July 24.
– 2023: The fieldwork for this cycle was carried out from May 31 to July 24.

This thorough, multi-year survey approach not only enriches our analysis by virtue of its vast dataset but also provides a nuanced understanding of shifting public sentiments across different political timelines. The descriptive statistics for the entire sample (spanning seven years) are presented in Table 3.6.

3.2.2 Consistency Analysis

First, we conducted a consistency analysis to determine the relevance of the created factors. We used Cronbach's Alpha to assess whether the variables derived from the literature were consistent. In cases where eliminating an item would only marginally improve the overall index, we chose to retain the variable to preserve its information. For instance, in the Civic Use factor (Table 3.7), removing the use7 variable would increase α from 0.86 to 0.88, but we decided to keep it (the full process is detailed in Appendices A.1 and A.2). Regarding social media use typology, we opted to work with two factors: civic use (encompassing political and informational use) and other use (covering social interaction and entertainment). As Table 3.7 demonstrates, both factors showed consistency.

Table 3.7 Dimensions, factors, variables, and results of the reliability analysis

Dimension	Factor	Variable	α if the item was deleted	α
Type of social media use	Civic use	use4	0.82	0.86
		use5	0.78	
		use6	0.82	
		use7	0.88	
	Non-civic use	use1	0.59	0.68
		use2	0.62	
		use3	0.56	
Political attitudes	Internal efficacy	inef2	0.72	0.67
		inef4	0.47	
		inef5	0.51	
	External efficacy	exef1	0.59	0.66
		exef3	0.52	
		exef4	0.58	
	Online political efficacy	ope1	0.78	0.83
		ope2	0.75	
		ope3	0.80	
		ope4	0.81	
	Political interest	polint1	0.89	0.87
		polint2	0.82	
		polint3	0.81	
		polint4	0.81	
		polint5	0.84	

Note This is an example considering all series from 2017, 2018, 2019, 2020, 2021, 2022, and 2023. The decision on which observable variable to retain may vary depending on the series used. For Online Political Efficacy, all series were used except for 2017, as the set of questions was included starting in 2018

In the case of political attitudes, the results were different from those mentioned above. For internal efficacy, the consistency analysis suggested we keep variables inef2, inef4, and inef5 and eliminate variables inef1 and inef3. The negative phrasing utilized in some questions is a possible explanation for this result. As shown in Table 3.1, variables inef1 and inef3 are formulated in negative terms. Although the scale was inverted in the coding stage so that all variables were measured in the same way, Cronbach's Alpha suggested we eliminate them anyway. For external efficacy, the consistency analysis suggested we keep variables exef1, exef3, and exef4, and eliminate variable exef2. The explanation for this result could be similar to that for internal efficacy. The phrasing utilized was in positive terms only for variable exef2.

3.2 The Analysis 49

Although the other three variables were recoded, Cronbach's Alpha suggested we eliminate them. In the case of political interest, all variables were included, achieving a consistent general index.

3.2.3 Structural Equation Models (SEM)

In order to estimate the factors and their interactions, we use a system of structural equation models (SEM). In recent years, this methodology has been recommended for estimating the relationships between the Internet and engagement (Kim, 2020; Purdy, 2017). This technique can be divided into two parts: construction of the measurement model and the structural model. In this section, we will explain the construction of the measurement models, which allow us to create the latent variables that will be used in our analysis. In the subsequent section (Chap. 4—Results and Discussion), we will focus on the structural model, incorporating the latent variables while also exploring their relationships with other relevant observable variables.

In the analysis of latent variables, it is crucial to distinguish between two fundamental types: reflective and formative latent variables (Tang et al., 2020). Our discussion thus far has primarily focused on reflective latent variables, which are theoretical constructs manifested through observable indicators. In this model, the latent variable is considered the cause of these indicators. However, it is equally important to consider formative latent variables. Unlike reflective variables, formative latent variables are constructed from their indicators, which are considered causal to the latent construct. This distinction is not merely semantic; it has significant implications for model specification, reliability and validity assessment, and the interpretation of causal relationships.

As shown in Figs. 3.1, 3.2, 3.3, and 3.4, the measurement model involves the creation of five factors, considering the results obtained in the consistency analysis already shown in the previous section. This technique is especially relevant when measuring internal efficacy and external efficacy. In some instances, we decided to create an aggregate variable by adding the values of each response or by simply calculating the average of each efficacy (Hansen & Tue Pedersen, 2014; Stoycheff et al., 2016; Zhang et al., 2010). Nonetheless, the literature suggests using structural equation modeling because it best captures the theoretical treatment of the concept (Bollen et al., 2008).

Political attitudes are usually considered when explaining the political or informational use of social media. However, many studies have encountered theoretical and methodological problems when conceptualizing and measuring the variables, which may be due to the lack of data and the use of predesigned questionnaires. These problems are notorious in the case of political efficacy, a concept that has been studied in a dissimilar way, and not always following the suggestions found in the literature. In this chapter, we propose a methodology that would recover the three-dimensionality of political efficacy (including online political efficacy) and also facilitate the creation of factors based on different questions, as suggested in the

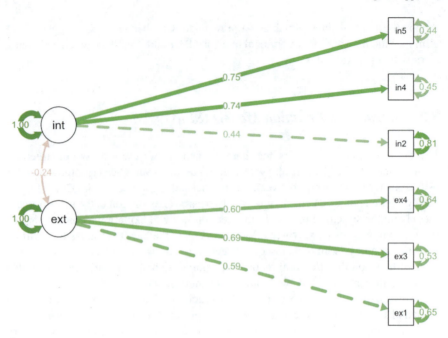

Fig. 3.1 Measuring model for political efficacies

literature. As mentioned, the use of structural equation models (SEM) has recently been suggested for researching the relationships between the Internet and engagement because it allows us to create factors commonly studied in the field of public communication and estimate their relationships simultaneously. We believe that the consistency and robustness of the models presented in this chapter provide a relevant contribution to the study on the potential democratizing power of digital platforms.

The strength of SEM lies in its ability to construct a latent variable from measurements of associated variables, which can then be used in a model to explain causal relationships. In this case, Figs. 3.1, 3.2, 3.3, and 3.4 illustrate that the causal arrows flow from the latent variables—political efficacies, social media use, and political interest—toward the variables measured through the previously described questions. The rationale behind this causality is that the common variance among these variables is produced by the latent variable we aim to measure. From this common variance, we construct the variable that is presumed to cause these variations in the observed measurements.

3.2 The Analysis

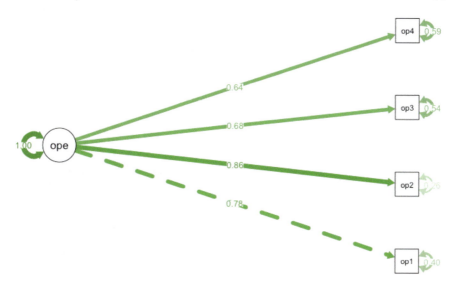

Fig. 3.2 Measuring model for online political efficacy (2018–2023)

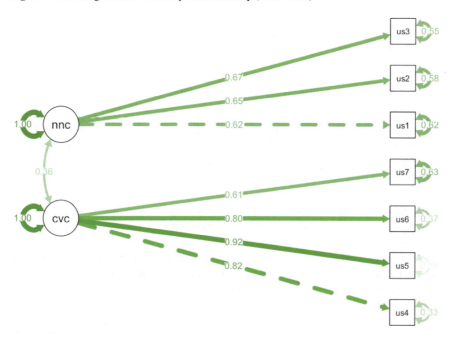

Fig. 3.3 Measuring model for social media use

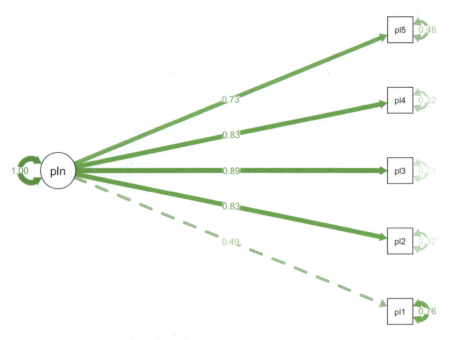

Fig. 3.4 Measuring model for political interest

3.2.4 Considering the Territorial Context

Historically, political attitudes have been primarily analyzed through individual factors or institutional elements tailored to each study. Yet, territorial and contextual elements haven't consistently been accorded the same degree of focus.

Recent developments in the "Geography of Discontent" literature illuminate a new perspective. They suggest that certain feelings of discontent and unrest might emerge from specific territories that perceive themselves as marginalized. Depending on the context, these territories may bear distinct characteristics.

In this context, recent contributions invite us to rethink the impact of territorial context on civic engagement. These contributions transcend traditional studies of the rural-urban divide and incorporate new approaches based on trajectories, feelings, and experiences in different regions, cities, or neighborhoods.

3.2 The Analysis

3.2.4.1 Center-Periphery Cleavage

While the notion of territory is by no means novel in political science, the methodological approaches to discern the spatial patterns of political unease can be manifold.

As previously mentioned, our study draws on a subnational sample, providing a localized representation that encompasses 10 specific municipalities. These municipalities collectively account for 70% of the region's total population. It's crucial to note that our sample strictly pertains to the urban zones within these municipalities; thus, respondents from rural areas are not included. Given these sample constraints, one of our initial exploratory lenses was the classic center-periphery cleavage.

In our specific context, three municipalities can be categorized under the political "center",[1] whereas seven fall within the political "periphery". This classification arises from the dominant intraregional centralization observed in the region. The past decade has seen growing tensions, especially among peripheral municipalities, some of which have expressed desires to break away from Valparaíso and establish a new region, tentatively termed the "Aconcagua region".

While Fig. 3.5 illustrates the geographical distribution of these cities, Table 3.8 presents some demographic characteristics, specifically the population and its percentage of the total population in the region. As we can see, with the exception of Concón—an affluent area—the peripheral municipalities tend to be smaller in population.

Despite this, it is important to note that the distinction between center and periphery does not necessarily correlate with other socioeconomic factors. These inequities are more related to political, historical, and cultural elements, marked by the intraregional centralism prevalent in Chile. Typically, residents of the periphery report that political decisions affecting them are made either from Santiago (the national capital) or Valparaíso (the regional capital), leaving them marginalized from subnational affairs.

Figures 3.6, 3.7, and 3.8 display indicators of poverty, employment, and population for each of the ten cities included in the sample. Although there is some territorial heterogeneity, these indicators do not clearly differentiate between the three cities that make up the center and the others. For example, Concón (center) has low poverty rates, unlike Valparaíso (center). Conversely, Los Andes (periphery) has high employment rates, unlike Quilpué (periphery).

To encapsulate, in the Chilean context, it is pertinent to analyze political attitudes through the center-periphery dichotomy, reflecting the inherent territorial and political dynamics at play.

---

[1] In addition to the Viña del Mar-Valparaíso conurbation (regional capital), the city of Concón was also considered as a political center. This latter area is adjacent to the main conurbation and has attracted affluent segments in recent decades.

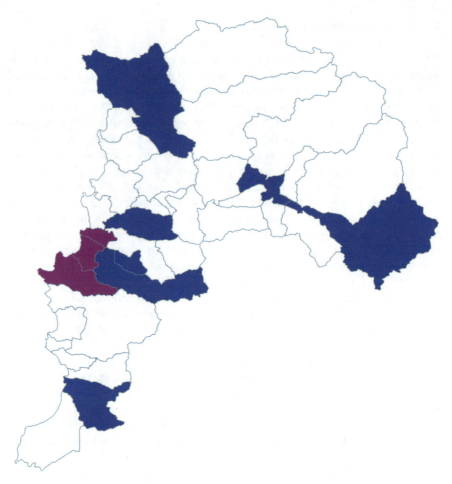

Fig. 3.5 Political Center of Valparaíso

3.2.4.2 Neighborhoods That Don't Matter

A recent strand within the geography of discontent literature offers different approaches, especially in the context of Europe and the United States. It underscores regions that have experienced a trajectory of decline, emanating a sense of being left behind. Interestingly, these are not necessarily impoverished or peripheral regions; instead, they are areas that once flourished but now seem neglected.

This notion is often associated with territories that have undergone deindustrialization, particularly in the United States (Rodríguez-Pose et al., 2021), or are ensnared in what some scholars refer to as the "development trap" (Diemer et al., 2022; Rodríguez-Pose et al., 2024), especially in Europe (Rodríguez-Pose et al., 2023). Yet, transplanting these ideas directly to Latin American contexts, or Chile

3.2 The Analysis

Table 3.8 Center/periphery cleavage

Municipality	Municipal population	% of regional population	Classification
Los Andes	66708	3.67	Periphery
La Ligua	35390	1.95	Periphery
Quillota	90517	4.98	Periphery
San Antonio	91350	5.03	Periphery
San Felipe	76844	4.23	Periphery
Concón	42152	2.32	Center
Valparaíso	296655	16.34	Center
Viña del Mar	334248	18.41	Center
Quilpué	151708	8.35	Periphery
Villa Alemana	126548	6.97	Periphery
Total		100	

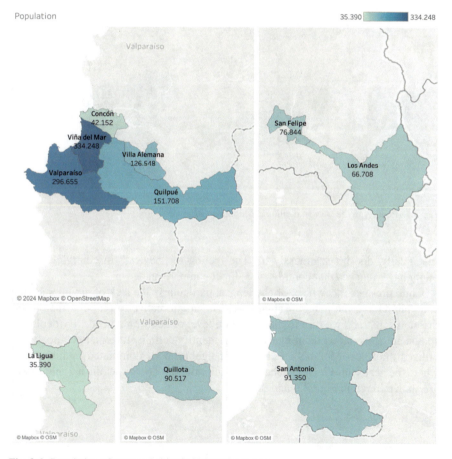

Fig. 3.6 Population of surveyed cities in Valparaíso Region

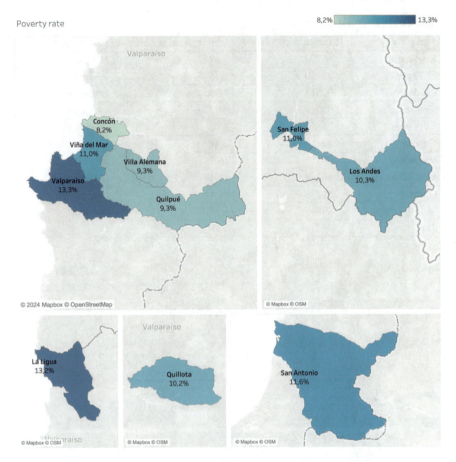

Fig. 3.7 Poverty in the surveyed cities in Valparaíso Region

specifically, doesn't seamlessly fit. The feeling of "neglect" in emerging nations might have nuances that complicate a direct comparison.

Chile presents its own set of challenges, particularly the lack of granular socioeconomic data at the neighborhood level. As an alternative to pinpoint the "places that don't matter", we turn to the education level of household heads. By deploying spatial autocorrelation techniques, we can identify certain clusters. Specifically, we utilized the Local Moran's I to earmark neighborhoods dominated by households where the heads have lower educational backgrounds. These specific neighborhoods are shown in red in Figs. 3.9 and 3.10.

3.2 The Analysis

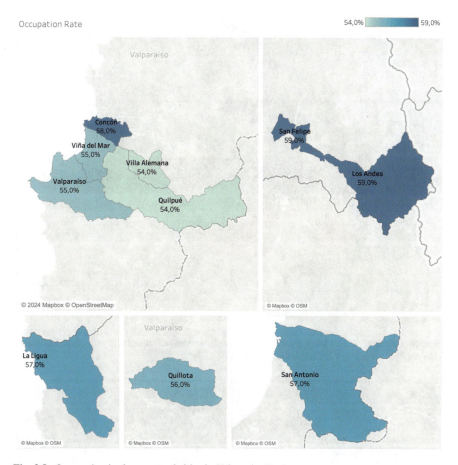

Fig. 3.8 Occupation in the surveyed cities in Valparaíso Region

A crucial distinction to bear in mind is that these identified areas don't inherently signify vulnerability. In Chile, and much of Latin America, vulnerability is frequently linked to informal settlements or "slums". Despite their material deficits, these settlements do not necessarily experience neglect. In contrast, informal settlements in Chile often receive regular governmental assistance, support from NGOs, and aid from broader society. These areas, marked in yellow in Figs. 3.9 and 3.10, sharply contrast with the neighborhoods shown in red, which are characterized by households headed by individuals with lower educational backgrounds. This differentiation aligns with the notion of "places that don't matter", suggesting areas that may not be impoverished but feel overlooked or undervalued.

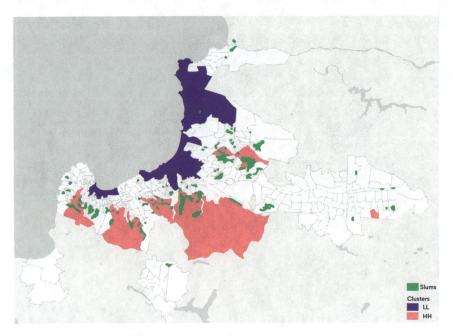

Fig. 3.9 Concentration of low education (in red) and informal settlements (in yellow) in Valparaíso, Viña del Mar, Concón, Villa Alemana, and Quilpué

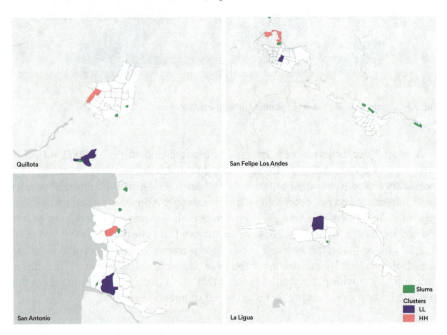

Fig. 3.10 Concentration of low education (in red) and informal settlements (in yellow) in San Felipe, Los Andes, Quillota, San Antonio, and La Ligua

References

Balch, G. I. (1974). Multiple indicators in survey research: The concept "Sense of Political Efficacy". *Political Methodology,1*(2), 1–43. http://www.jstor.org/stable/25791375

Bollen, K. A., Rabe-Kesketh, S., & Skrondal, A. (2008). Structural equation models. In D. Collier & J. M. Box-Steffensmeier (Eds.), *The Oxford handbook of political methodology* (pp. 432 455). Oxford University Press.

Campbell, A., Gurin, G., & Miller, W. E. (1954). *The voter decides*. Row, Peterson & Co.

Chan, M., Chen, H.-T., & Lee, F. L. F. (2016). Examining the roles of mobile and social media in political participation: A cross-national analysis of three Asian societies using a communication mediation approach. *New Media & Society,19*(12), 2003–2021. https://doi.org/10.1177/1461444816653190

Coverse, P. E. (1972). Change in the American electorate. In A. Campbell & P. E. Converse (Eds.), *The human meaning of social change*. Russell Sage Foundation.

Diehl, T., Weeks, B. E., & Gil de Zúñiga, H. (2016). Political persuasion on social media: Tracing direct and indirect effects of news use and social interaction. *New Media & Society,18*(9), 1875–1895. https://doi.org/10.1177/1461444815616224

Diemer, A., Iammarino, S., Rodríguez-Pose, A., & Storper, M. (2022). The regional development trap in Europe. *Economic Geography*, 1–23. https://doi.org/10.1080/00130095.2022.2080655

Gil de Zúñiga, H., Molyneux, L., & Zheng, P. (2014). Social media, political expression, and political participation: Panel analysis of lagged and concurrent relationships. *Journal of Communication,64*(4), 612–634. https://doi.org/10.1111/jcom.12103

Hansen, K. M., & Tue Pedersen, R. (2014). Campaigns matter: How voters became knowledgeable and efficacious during election campaigns. *Political Communication,31*(2), 303–324. https://doi.org/10.1080/10584609.2013.815296

Hyun, K. D., & Kim, J. (2015). Differential and interactive influences on political participation by different types of news activities and political conversation through social media. *Computers in Human Behavior,45*, 328–334. https://doi.org/10.1016/j.chb.2014.12.031

Kim, J. W., Park, S., & Sang, Y. (2020). A longitudinal study on the effects of motivational mobile application uses on online news engagement. *International Journal of Mobile Communications,18*(3), 327–342. https://doi.org/10.1504/IJMC.2020.107103

Lupia, A., & Philpot, T. S. (2005). Views from inside the net: How websites affect young adults' political interest. *The Journal of Politics,67*(4), 1122–1142. https://doi.org/10.1111/j.1468-2508.2005.00353.x

Oskarson, M. (2007). Social risk, policy dissatisfaction and political alienation: A comparison of six European countries. In S. Svallfors (Ed.), *The political sociology of the welfare state: Institutions, social cleavages and orientations*. Stanford University Press.

Purdy, S. J. (2017). Internet use and civic engagement: A structural equation approach. *Computers in Human Behavior,71*, 318–326. https://doi.org/10.1016/j.chb.2017.02.011

Rodríguez-Pose, A., Dijkstra, L., & Poelman, H. (2024). The geography of EU discontent and the regional development trap. *Economic Geography,100*(3), 213–245. https://doi.org/10.1080/00130095.2024.2337657

Rodríguez-Pose, A., Lee, N., & Lipp, C. (2021). Golfing with Trump. Social capital, decline, inequality, and the rise of populism in the US. *Cambridge Journal of Regions, Economy and Society,14*(3), 457–481. https://doi.org/10.1093/cjres/rsab026

Rodríguez-Pose, A., Terrero-Dávila, J., & Lee, N. (2023). Left-behind versus unequal places: Interpersonal inequality, economic decline and the rise of populism in the USA and Europe. *Journal of Economic Geography,23*(5), 951–977. https://doi.org/10.1093/jeg/lbad005

Stoycheff, E., Nisbet, E. C., & Epstein, D. (2016). Differential effects of capital-enhancing and recreational Internet use on citizens' demand for democracy. *Communication Research*, 0093650216644645. https://doi.org/10.1177/0093650216644645

Strömbäck, J., & Shehata, A. (2010). Media malaise or a virtuous circle? Exploring the causal relationships between news media exposure, political news attention and political interest. *European Journal of Political Research,49*(5), 575–597. https://doi.org/10.1111/j.1475-6765.2009.01913.x

Tang, J., Ren, H., & Folmer, H. (2020). Chapter 5: Subjective wellbeing as valuation system of environmental quality: An environmental social sciences approach. In D. Maddison, K. Rehdanz, & H. Weolsch (Eds.), *Handbook on wellbeing*. Happiness and the Environment: Elgar.

Torcal, M. (2006). Desafección institucional e historia democrática en las nuevas democracias. *Revista SAAP, 2*(3), 591–634.

Yang, H."C."., & DeHart, J. L. (2016). *Social media use and online political participation among college students during the US election 2012* (pp. 1–18). January–March: Social Media + Society.

Zhang, W., Johnson, T. J., Seltzer, T., & Bichard, S. L. (2010). The revolution will be networked: The influence of social networking sites on political attitudes and behavior. *Social Science Computer Review,28*(1), 75–92. https://doi.org/10.1177/0894439309335162

Chapter 4
Results and Discussion

Abstract In this chapter, we delve into the multifaceted dimensions of the digital divide, the implications of territorial dynamics, and the development of political discontent. Initial findings suggest that digital inequalities persistently shape Online Political Efficacy. This new attitude evaluates whether the digital world makes political systems more accessible, arguing that it provides a fresh perspective on the democratizing role of the Internet. Interestingly, territorial dimensions play a significant role: those further from political centers feel less competent in public affairs but more empowered by the Internet. Subsequently, we explore the geography of discontent, focusing on sentiments like abandonment, frustration, and resentment. Through the lens of External Efficacy, understood as the perception of system responsiveness, we unearth insights about feelings of political neglect, especially in "places that don't matter". Our empirical results demonstrate that regions marked by lower household head educational attainment exhibit diminished faith in the political system's responsiveness. Intriguingly, individual socioeconomic factors manifest in a counterintuitive manner, underscoring the need for a holistic, spatial approach to studying discontent. Finally, we share some initial insights into the study of political frustration and resentment. In sum, this chapter provides a nuanced understanding of how digital divides, territorial disparities, and emotive politics intersect in the realm of political behavior.

Keywords Geography of discontent · Political attitudes · Chile · Territorial inequalities · Digital inequalities

4.1 Political Attitudes and Digital Platforms

In today's world, understanding societal discontent without acknowledging the digital public sphere would be an oversight. Specifically, assessing the impact of digital media on political engagement is paramount. While much has been written on this topic, our discussion here zeroes in on the influence of social media platforms on the growth of political disaffection, a distinct manifestation of such discontent.

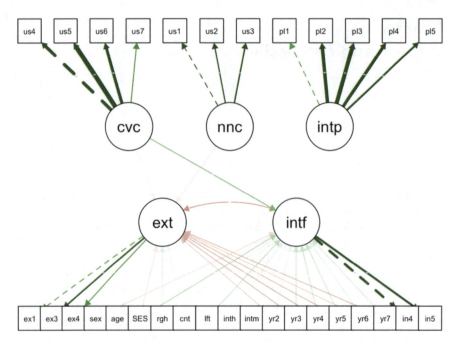

Fig. 4.1 Structural equation model for classical political efficacy. *Note* These results are based on Fierro et al. (2023b). In this example, data from 2017 to 2023 were included

As displayed in Fig. 4.1 and Table 4.1, our findings, based on a preliminary study by Fierro et al. (2023a), underline the need for a nuanced approach when analyzing the influence of social media on political attitudes related to disaffection. The civic use of digital platforms—which encompasses both informational and political usage—correlates positively with these attitudes. In essence, individuals who engage with social media for civic purposes tend to feel more politically competent and believe that the political system is responsive to them.

However, this relationship between social media usage and political disaffection shifts when considering non-civic use—a category that comprises recreational and social engagement. Here, the connection between social media use and external efficacy becomes negative. To elaborate, those who primarily utilize social media for entertainment or socializing are more likely to feel that the political system doesn't address their needs, reflecting a heightened sense of political disaffection.

These insights compel us to reassess the role social networks play in fostering disaffection. The differential usage of these platforms may be exacerbating the already-existing chasm between those contented with the system and those disillusioned by it.

Beyond these findings, the significance of other determinants is worth noting. For instance, our data indicates that women are less likely to perceive themselves as politically competent. Similarly, the influence of socioeconomic status is intriguing.

4.1 Political Attitudes and Digital Platforms

Table 4.1 Structural equation model for classical political efficacies

		Internal efficacy	External efficacy
Demographics	Gender	−0.156***	−0.021
		(0.021)	(0.017)
	Age	0.001	−0.000
		(0.001)	(0.001)
	SES	0.138***	−0.039***
		(0.011)	(0.009)
Ideology	Right	0.100**	0.118***
		(0.032)	(0.026)
	Center	0.091**	0.063*
		(0.035)	(0.029)
	Left	0.259***	0.034
		(0.028)	(0.023)
Access to Internet	Home	0.101**	0.002
		(0.029)	(0.024)
	Mobile	0.074*	−0.041
		(0.038)	(0.031)
Use of social media	Civic	0.174***	0.015*
		(0.008)	(0.006)
	Non-civic	−0.001	−0.031**
		(0.013)	(0.011)
Year	2018	0.130**	−0.238***
		(0.043)	(0.036)
	2019	0.230***	−0.316***
		(0.041)	(0.034)
	2020	0.257***	−0.345***
		(0.041)	(0.035)
	2021	0.234***	−0.316***
		(0.041)	(0.035)
	2022	0.114**	−0.223***
		(0.041)	(0.034)
	2023	0.193***	−0.312***
		(0.041)	(0.035)

Note Standard errors in (). p-value < 0.05*; p-value < 0.005**; p-value < 0.001***. These results are based on Fierro et al. (2023b). In this example, data from 2017 to 2023 were included

Individuals at lower socioeconomic levels tend to feel less politically competent; however, they simultaneously feel that the system addresses their needs. Additionally, political alignment—whether leaning left or right—positively correlates with both forms of efficacy, suggesting a stronger political connection.

4.2 Territorial Perspective on the Digital Divide

In the literature review, we introduced the concept of the digital divide, which delineates the disparities between those who harness the benefits of the Internet and the digital realm and those who don't.

There are a myriad of approaches to studying the digital divide and digital inequalities, all of which offer unique and invaluable insights. As mentioned earlier, recent discourse has advanced the notion of "Online Political Efficacy", a concept tailored to capture the potential political empowerment facilitated by digital platforms. Specifically, this novel attitude seeks to gauge whether the political system becomes more accessible and responsive via the Internet, thereby enhancing the value of political action.

With the lens of Online Political Efficacy, it becomes feasible to reevaluate the longstanding debate concerning the Internet's democratizing role, allowing for the inclusion of a territorial component.

Using the central-peripheral cleavage as a backdrop, Fig. 4.2 and Table 4.2 display results from a Structural Equation Model gauging all forms of political efficacy—Internal, External, and Online (Fierro et al. 2023a). Notably, these results indicate

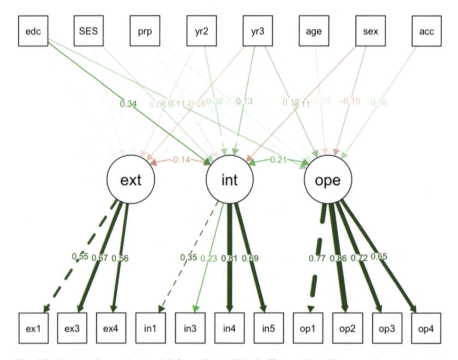

Fig. 4.2 Structural equation model for online political efficacy. *Note* These results are based on Fierro et al. (2023a). Some results may vary slightly due to the use of different software and specifications

4.2 Territorial Perspective on the Digital Divide

Table 4.2 Structural equation model for classical and online political efficacies

		Int. Eff.	Ext. Eff.	Online Pol. Eff.
Territory	Periphery	−0.031*	0.026	0.074*
		(0.016)	(0.025)	(0.031)
Demographics	Sex	−0.104***		−0.186***
		(0.016)		(0.032)
	Age		0.001	−0.003**
			(0.001)	(0.001)
	Education	0.062***	−0.007	0.043***
		(0.006)	(0.008)	(0.010)
	SES	0.033**	−0.017	0.025
		(0.011)	(0.017)	(0.022)
Access to Internet	General access			0.079***
				(0.019)
Year	2019	0.079***	−0.063**	0.049
		(0.019)	(0.031)	(0.038)
	2020	0.128***	−0.103*	0.239***
		(0.020)	(0.031)	(0.039)

Note Standard errors in (). p-value < 0.05*; p-value < 0.005**; p-value < 0.001***. These results are based on Fierro et al. (2023a). Some results may vary slightly due to the use of different software and specifications

that residing in the political periphery correlates with diminished internal efficacy. This suggests that individuals geographically distant from the political epicenter tend to feel less equipped to participate in public affairs, culminating in increased political disaffection as per our definitions. However, these very individuals also perceive that the Internet equips them with enhanced understanding and influence over the system.

These findings spotlight the unique territorial dynamics underscored by intra-regional centralization. Within a context where most developments are concentrated in the regional capital, the Internet appears as a beacon of integration for areas that are politically marginalized. Building upon this observation, we can revisit the perennial debate: Does the Internet serve as a unifying tool, or does it merely amplify existing power dynamics? While evidence from the preceding section leans toward the latter, the introduction of a territorial perspective suggests the situation is more nuanced than previously thought.

4.3 Places That Don't Matter and the Feeling of Abandonment

In our earlier discussions, we introduced the concept of External Efficacy as the perception of system responsiveness. This pertains to an individual's belief that the political system is receptive and responsive to citizen demands. Interestingly, this definition aligns well with the literature on the geography of discontent, which often relates to feelings of abandonment. Therefore, we could infer that citizens residing in the so-called "places that don't matter" would, theoretically, exhibit lower levels of external efficacy.

The connection between territorial characteristics and feelings of abandonment, although frequently cited in recent literature, has not been comprehensively explored in empirical studies. The ensuing results delve into this relationship, offering an attitudinal perspective on the geography of discontent.

To further our analysis, we identified these "places that don't matter" as areas with a concentration of families whose household heads have a low educational attainment. As previously elucidated, given the specificities of our case study, this serves as a useful criterion for pinpointing places that have been overlooked, even if they are not necessarily the most vulnerable.

The results, presented in Table 4.3, are both robust and revealing. In fact, inhabitants of areas where families with household heads of lower educational attainment

Table 4.3 Structural equation model for external efficacy

		EPE
Demographics	Age	0.037*
		(0.019)
	SES	−0.57*
		(0.023)
	Gender	0.031†
		(0.018)
	Education (individual)	−0.036
		(0.024)
Territorial	Periphery	0.057**
		(0.019)
	Extreme poverty	−0.01
		(0.019)
Neighborhood	Low education	−0.056**
		(0.018)
Year	2020	−0.023
		(0.021)
	2021	0.019
		(0.021)

Note Standard errors in (). p-value < 0.05*; p-value < 0.005**; p-value < 0.001***; p-value < 0.1†. These results are reported in Fierro et al. (2024)

reside tend to report lower levels of external efficacy. In other words, they are more likely to believe that the system does not respond to their needs. Intriguingly, the individual's socioeconomic status exhibits an inverse relationship. As the socioeconomic level decreases, the levels of external efficacy rise. Moreover, another point of interest is that, contrary to the territorial variable of the household head's education, the individual education level of the respondent does not significantly correlate with the perception of system responsiveness in our model.

We believe that these findings underscore the importance of analyzing discontent not merely from an individual standpoint but also in the context of geographic spaces.

4.4 Preliminary Results About Resentment and Frustration

In our literature review, we delved into the geography of discontent, which is replete with discussions on feelings of frustration, rage, abandonment, and other emotions that arise from residing in specific locales with distinctive characteristics and histories. Interestingly, while these studies explore discontent through voting patterns, they seldom dive deep into the realm of attitudinal elements.

By utilizing the concept of external efficacy, we've made strides in pinpointing a particular sentiment linked to feelings of abandonment and being overlooked by the political system.

Nevertheless, through our exploration of political disaffection—especially in the context of political efficacies—it's plausible to venture into two other paramount political sentiments: frustration and resentment.

In the annals of political attitudes and behavior, consensus remains elusive concerning how best to approach these emotions. It's evident, however, that issues of temporality and expectation come into play. Analogous to feelings of abandonment, political frustration doesn't solely stem from absolute deprivation or material deficiencies.

The results previously obtained using the SEM measurement model to generate the external and internal efficacy variables allow us to calculate new concepts. This is possible because we used the same variables and survey respondents to estimate both efficacy variables. Consequently, this makes them comparable across the years studied and enables us to generate measurements for concepts that are otherwise difficult to quantify.

Taking into account the matter of expectations, Gamson's classic hypothesis sheds light on this intricacy. Reinterpreted by Craig (1990), a scenario characterized by high internal efficacy but low external efficacy emerges as a fertile ground for informal participation and acts of protest. Consequently, we'll leverage this reinterpretation to discern an initial sensation of frustration, perceived as the gap between external and internal efficacy.

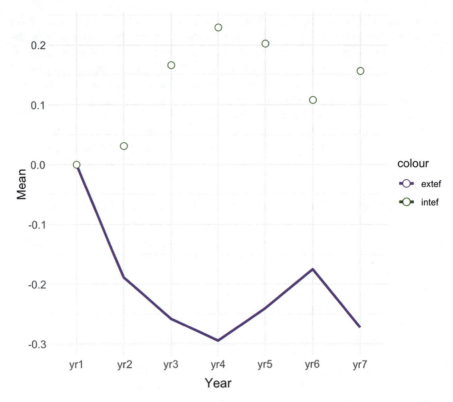

Fig. 4.3 Gamson's hypothesis. *Note* Circles represent the means for internal efficacy, while the line represents the means for external efficacy. These results are based on Aroca et al. (2024). In this example, data from 2017 to 2023 were included

Figure 4.3 showcases preliminary findings on this gap. The graphical representation is rather straightforward, particularly when considering the tumultuous social unrest witnessed in Chile between late 2019 and early 2020. It was precisely during this period that the disparity between the efficacies was most pronounced.

Nevertheless, as we have contended, it is not solely about expectations; temporality also plays a pivotal role. So, beyond Gamson's hypothesis, to quantify frustration and resentment—two salient emotions in discontent geography literature—we can commence with a basic distinction between flow and stock variables. Following this trajectory, the forthcoming proposal, elaborated upon in Aroca et al. (2024), perceives frustration as a flow variable and resentment as a stock variable, which accumulates the former.

4.4 Preliminary Results About Resentment and Frustration

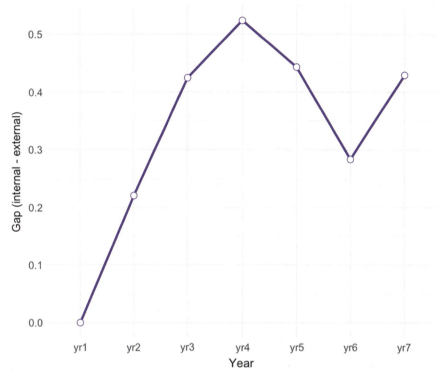

Fig. 4.4 Evolution of the gap (frustration). *Note* These results are based on Aroca et al. (2024). In this example, data from 2017 to 2023 were included

Figures 4.4 and 4.5 delineate the trajectory of both sentiments over the considered seven-year span. Under this framework, the intriguing aspect is that resentment, being a stock variable, continues to amass as long as frustration remains positive. This implies that even if feelings of frustration have been on a decline since 2020, resentment continues to pile up.

In 2023, we observe a renewed increase in frustration, marking a change in the trend compared to previous years. This indicates that resentment continues to grow, but at a higher rate than in preceding years. Such a trajectory could potentially lead to another social upheaval if these levels reach a point where the population loses faith in the current institutional channels' ability to meet their expectations.

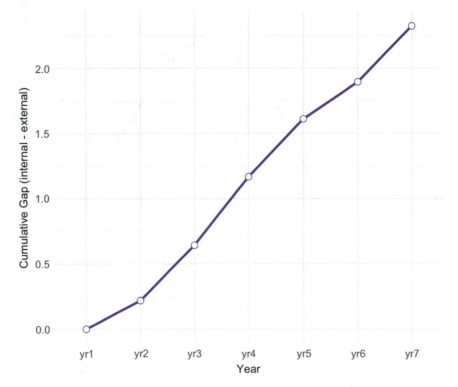

Fig. 4.5 Accumulation of the gap (resentment). *Note* These results are based on Aroca et al. (2024). In this example, data from 2017 to 2023 were included

References

Aroca, P., Fierro, P., & Sánchez-Barría, F. (2024). Frustration and resentment in politics: An application to the Chilean case. SocArXiv https://doi.org/10.31235/osf.io/ng965

Craig, S. C., Niemi, R. G., & Silver, G. E. (1990). Political efficacy and trust: A report on the NES pilot study items. *Political Behavior, 12*(3), 289–314. http://www.jstor.org/stable/586303

Fierro, P., Aroca, P., & Navia, P. (2023a). The center-periphery cleavage and online political efficacy (OPE): Territorial and democratic divide in Chile, 2018–2020. *New Media & Society, 25*(6), 1335–1353. https://doi.org/10.1177/14614448211019303

Fierro, P., Aroca, P., & Navia, P. (2023b). Political disaffection in the digital age: The use of social media and the gap in internal and external efficacy. *Social Science Computer Review, 41*(5), 1857–1876. https://doi.org/10.1177/08944393221087940

Fierro, P., Aravena-Gonzalez, I., Aroca, P., & Rowe, F. (2024). Geographies of discontent: Measuring and understanding the feeling of abandonment in the Chilean region of Valparaiso (2019–2021). *Cambridge Journal of Regions, Economy and Society,* rsae004. https://doi.org/10.1093/cjres/rsae004

Chapter 5
Conclusion

Abstract This book aims to explore the intricate relationship between spatial inequalities, political discontent, and the potential role of the Internet in including inhabitants of politically marginalized areas. Focusing on Chile's Valparaíso region, we provide a theoretical and methodological framework to understand the anger, frustration, and sense of abandonment experienced by citizens. Our study highlights that discontent is not solely a result of deprivation and poverty but also of poor relationships with public services and persistent marginalization, leading to violence and further isolation. We employed a survey dataset and structural equation models to analyze these phenomena, considering both territorial context and temporal dimension. Despite significant contributions, our work faces limitations such as the generalization of external efficacy measures and the cross-sectional nature of our data. Future research should address these gaps by exploring more specific relationships with various administrative levels and employing longitudinal data. The discussion emphasizes the cyclical nature of abandonment and its consequences in Chile, where social unrest led to the destruction of infrastructure in marginalized areas. The role of the Internet in addressing these issues is complex, requiring attention to digital inequities and challenges like misinformation and polarization. Our findings underscore the need for a comprehensive approach to integrating marginalized communities into the political process.

Keywords Geography of discontent · Political attitudes · Chile · Territorial inequalities · Digital inequalities

5.1 Final Remarks

5.1.1 Conclusion

The objective of this book is to propose a theoretical and methodological framework to address two primary phenomena: the relationship between space and political discontent, and the potential role of the Internet in including inhabitants of politically marginalized areas. To achieve this, we have focused on a specific case study and various practical issues encountered in analyzing these phenomena.

Our framework is grounded in connecting contributions from various disciplines, essential for understanding the studied phenomena. We based our initial analysis on the concepts of political interest and political efficacy, which, for Latin America—including Chile, our case study—have been closely associated with political disaffection. We extended our analysis to the concept of online political efficacy, applying the sensations of efficacy to the online world.

We hypothesize that political efficacy—in its three dimensions: internal, external, and online—can be a useful concept for understanding the sense of abandonment experienced by inhabitants of left-behind places, warranting a revisit.

This book's core contribution is proposing a methodology to measure and utilize these attitudes. We suggest a specific questionnaire and the understanding of these attitudes as latent variables that respond to observed variables and interact with other elements. Using Structural Equation Models (SEM), widely promoted in political psychology literature, we analyze sensations, emotions, and attitudes.

After that, we apply this proposal to the specific scenario of our case study, the Chilean region of Valparaíso. In our view, the results are compelling and extend our understanding of the two phenomena in question.

First, there is a relationship between political disaffection and the use of digital platforms, though perhaps not as typically believed. While political use of these tools positively impacts internal efficacy (self-competence beliefs) and external efficacy (system responsiveness beliefs), social or recreational use does not. In fact, social use is negatively related to external efficacy, suggesting that those who use digital platforms for non-informational or political purposes are more likely to feel that the political system does not respond to their interests, leading to greater disaffection. These results imply that the Internet might widen the gap between the politically engaged and disengaged, creating a vicious cycle of discontent.

However, the inclusion of territorial context changes the analysis. Using the concept of online political efficacy, we find that inhabitants of peripheral areas believe digital platforms provide greater opportunities to be heard, even if they feel less competent to participate in public affairs. These results suggest that territorial marginalization affects attitudes and internal resources, with peripheral residents feeling less competent due to their location. Yet, they also indicate that the digital divide takes on a new significance when territorial context is considered. Our initial individual-level analyses suggested a pessimistic view of the Internet's democratic role, but including territorial marginalization reveals that the Internet might be a useful tool for including residents of abandoned areas.

We also explored the use of classical efficacies to transition from the geography of voting to the geography of discontent, focusing on attitudes preceding voting behavior. Using spatial autocorrelation techniques, we identified neighborhoods in marginal conditions, finding that living in these left-behind areas is associated with lower external efficacy. These findings confirm a recurrent assumption in the geography of discontent literature: inequalities translate into a sense of abandonment, leading to support for anti-system narratives. Our focus is on this initial connection.

5.1 Final Remarks 73

Finally, we share preliminary results on studying political frustration and resentment, revisiting classical efficacies and applying Gamson's hypothesis. We observe that time matters—sensations accumulate and evolve. In our particular case study of Chile, these descriptive results, which show the evolution of frustration, become especially important for understanding the events of October 2019, when more than two million citizens took to the streets to protest various social demands. Indeed, that year witnessed the largest gap between internal and external political efficacy. Citizens felt more competent than ever—though this does not necessarily translate to objective knowledge—while their perceptions of system responsiveness reached an all-time low. This gap has been suggested as the optimal condition for seeking alternative mechanisms, which in the context of Latin America often translates to violence and protest. We believe that these temporal considerations of discontent open new avenues for research, allowing us to further extend our understanding of the geography of discontent.

5.1.2 Discussion

We began by suggesting that the study of discontent must incorporate both territorial context and temporal dimension and greater specificity when discussing unrest. This specificity is crucial for our proposal, as it leads us to consider various political attitudes that underpin electoral participation and influence other issues prevalent in certain Global South contexts, such as abstention and violence.

The nature of abandonment in Chile and Latin America is peculiar. As our findings suggest, abandonment does not necessarily correlate with deprivation and poverty but rather with poor relationships with space and public services and a persistent trajectory of political marginalization, resulting in low investment and urban decay. This situation can generate anger, as has been argued. The problem is that in Chile and Latin America, this anger does not necessarily channel through formal participation mechanisms but also through alternative outlets, seeking spaces for participation that often result in violence and further destruction, which, in turn, leads to greater isolation and marginalization.

Chile's case is emblematic in this regard. The social unrest of 2019, which saw more than two million people take to the streets across the country, quickly led to the destruction of basic infrastructure. In Santiago, the capital, the metro system was paralyzed, with 118 stations damaged, representing 87% of the total. The violence and destruction were not localized in affluent areas but rather in popular sectors, where supermarkets were burned, stores were looted, and a climate of insecurity and violence was established. The consequences of this abandonment form a cyclical pattern: inequities generate anger, which manifests in destruction, leading to greater spatial inequities.

The role of the Internet in including marginalized areas must be examined carefully, considering both its potential and the challenges it presents. During the pandemic, digital platforms sometimes amplified existing inequities. Online classes via

Zoom at universities illustrate this situation. Connections and environments were unequal, creating evident gaps. In physical classrooms, there was a relatively equal space where everyone interacted as peers. Digital classes disrupted this, with students unable to connect due to a lack of devices, poor Internet signals, and unsuitable study environments.

The challenge of digital inequities and their relationship with pre-existing inequities has been extensively studied in communication sciences (Helsper, 2012; Helsper & van Deursen, 2017; Helsper, 2021), necessitating a range of measures encompassing infrastructure, education, attitudes, and support networks. However, rethinking the Internet's role transcends the issue of inequities. For digital platforms to effectively integrate inhabitants of marginalized or distant areas from political centers, several digital life challenges must be addressed. Misinformation and homophily (often leading to both affective and ideological polarization) are just two examples that have become prominent challenges in democracies worldwide.

5.1.3 Limitations and Further Research

While our work significantly contributes to understanding the contextual elements that explain political discontent and the role of the Internet in including inhabitants of marginalized territories, it is essential to acknowledge certain limitations in our analysis. Some of these limitations were anticipated, while others were highlighted by reviewers and colleagues during academic conferences where we presented our findings.

For instance, this study measures the feeling of abandonment through the classic measure of external efficacy. Our questions assess the relationship between citizens and authorities/institutions without specifying which institutions are involved. This generic approach does not account for potentially significant differences. For example, the relationship between a resident of a neglected area and their local authority can be vastly different from their relationship with a national authority. These distinctions are crucial as they may vary greatly. In ongoing research, we have found that living in a peripheral city that feels insignificant within its region can affect political knowledge about regional issues but not necessarily about global matters. Similarly, residents of marginalized neighborhoods might have a distant relationship with their national or regional authorities, but this might not apply to more local representatives, such as neighborhood councils or mayors. Future research should explore these differences in depth, measuring the sense of responsiveness (external efficacy) in more specific terms, differentiating between various levels of administration and institutions.

Moreover, in the context of our study, it is pertinent to examine whether the latent variables we have utilized are reflective or formative in nature. This consideration can substantially influence our conclusions and the robustness of our model. For instance, in the field of environmental social sciences, Tang et al. (2020) have demonstrated how the conceptualization of subjective well-being as a valuation system of envi-

5.1 Final Remarks

ronmental quality can vary depending on whether it is considered a reflective or formative construct. The choice between reflective and formative models should be based on theoretical and empirical considerations. Reflective models assume that indicators are manifestations of the latent construct and, therefore, should be highly correlated. In contrast, formative models do not require indicators to be correlated, as they are considered components that "form" the latent construct. It is important to note that the decision to use a reflective or formative model has implications for estimation techniques and model evaluation criteria. For example, traditional confirmatory factor analysis techniques are more appropriate for reflective models, while variance-based structural equation modeling may be more suitable for formative models. In future research, it would be valuable to explore how the consideration of formative latent variables could enrich our analysis and provide a more comprehensive understanding of the phenomena under study. This could involve re-evaluating our theoretical constructs and considering alternative methods of data modeling and analysis. The work of Tang et al. (2020) provides a robust theoretical framework for conceptualizing and operationalizing these constructs in the context of environmental social sciences. Their approach offers valuable insights into the implications of choosing between reflective and formative models in environmental research. By expanding our consideration to include both reflective and formative latent variables, we can enhance the methodological rigor of our study and potentially uncover new insights into the complex relationships between our variables of interest.

Another significant limitation relates to the data used. Our argument emphasizes the importance of considering both space and time when studying discontent, which is reflected in our concepts of frustration and resentment. The idea of a stock variable—resentment—relies on tracking the two attitudes studied (internal and external efficacy) over time. Our project uses data collected over seven years (2017–2023), but the nature of the instrument is cross-sectional. There is no longitudinal tracking of the same individuals, only snapshots of specific moments with different respondents for each wave. To address this, we used pseudo-panel techniques, shifting from individual analysis to group analysis over time. However, it is crucial to explore how this proposal would apply at the individual level, as individuals (and residents of certain territories) experience frustration and resentment. Future projects could explore the possibility of collecting longitudinal data to test these ideas more thoroughly.

Additionally, another limitation is the geographic coverage of our study. The spirit of our project and this book is to share a theoretical and methodological proposal that allows us to explore the relationships between spatial context, discontent, and digital engagement. However, the application of this proposal in our case is limited to a specific area of Chile. For reasons outlined at the beginning of this book, we consider this case to be interesting and particularly useful—due to its political crisis, centralism, and high Internet penetration—but it is important to note that the findings cannot be generalized to the rest of the country or to the broader Latin American context. Our challenge is to extend our study to other contexts, which is difficult because data collection must have a specific local granularity and be representative of the territories studied.

Finally, in our work, we aimed to understand the impact of territorial context on the development of political attitudes. However, this remains far from fully grasping the broader spatial implications of discontent and engagement, particularly in relation to interaction, heterogeneity, and location. Future research should explore spatial spillovers by examining peer effects and investigating whether political attitudes extend beyond neighborhood boundaries. Addressing this challenge may also benefit from the use of spatial structural equation models (Folmer & Oud, 2008; Liu et al., 2005).

Given these considerations, this book should be read as a provocation that aims to connect some phenomena while simultaneously venturing into a concrete proposal for studying them. Each step certainly deserves continuous improvement.

References

Folmer, H., & Oud, J. (2008). How to get rid of W: A latent variables approach to modelling spatially lagged variables. *Environment and Planning A: Economy and Space, 40*(10), 2526–2538. https://doi.org/10.1068/a4078

Helsper, E. J. (2012). A corresponding fields model for the links between social and digital exclusion. *Communication Theory, 22*(4), 403–426. https://doi.org/10.1111/j.1468-2885.2012.01416.x

Helsper, E. J. (2021). *The digital disconnect: The social causes and consequences of digital inequalities*. Sage.

Helsper, E. J., & van Deursen, A. J. A. M. (2017). Do the rich get digitally richer? Quantity and quality of support for digital engagement. *Information, Communication & Society, 20*(5), 700–714. https://doi.org/10.1080/1369118X.2016.1203454

Liu, X., Wall, M. M., & Hodges, J. S. (2005). Generalized spatial structural equation models. *Biostatistics, 6*(4), 539–557. https://doi.org/10.1093/biostatistics/kxi026

Tang, J., Ren, H., & Folmer, H. (2020). Chapter 5: Subjective wellbeing as valuation system of environmental quality: An environmental social sciences approach. In D. Maddison, K. Rehdanz, & H. Weolsch (Eds.), *Handbook on wellbeing. Happiness and the environment*. Elgar.

Appendix

A.1 Reliability Test Political Efficacies

See Tables A.1 and A.2.

Table A.1 Reliability analysis for internal efficacy

Factor	Variable	α if the item was deleted	α
Internal efficacy	inef1	0.55	0.6
	inef2	0.60	
	inef3	0.61	
	inef4	0.46	
	inef5	0.51	
Internal efficacy	inef1	0.67	0.61
	inef2	0.59	
	inef4	0.41	
	inef5	0.45	
Internal efficacy	inef2	0.72	0.67
	inef4	0.47	
	inef5	0.51	

Note This is an example considering all series from 2017, 2018, 2019, 2020, 2021, 2022, and 2023. The decision on which observable variable to retain may vary depending on the series used

Table A.2 Reliability analysis for external efficacy

Factor	Variable	α if the item was deleted	α
External efficacy	exef1	0.38	0.55
	exef2	0.66	
	exef3	0.38	
	exef4	0.42	
External efficacy	exef1	0.59	0.66
	exef3	0.52	
	exef4	0.58	

Note This is an example considering all series from 2017, 2018, 2019, 2020, 2021, 2022, and 2023. The decision on which observable variable to retain may vary depending on the series used

Printed in the United States
by Baker & Taylor Publisher Services